CIVIL SURVEYING SAMPLE EXAMS

for the California Special Civil Engineer Examination

Second Edition

Peter R. Boniface, PhD, PLS

Professional Publications, Inc. • Belmont, CA

How to Locate and Report Errata for This Book

At Professional Publications, we do our best to bring you error-free books. But when errors do occur, we want to make sure you can view corrections and report any potential errors you find, so the errors cause as little confusion as possible.

A current list of known errata and other updates for this book is available on the PPI website at **www.ppi2pass.com/errata**. We update the errata page as often as necessary, so check in regularly. You will also find instructions for submitting suspected errata. We are grateful to every reader who takes the time to help us improve the quality of our books by pointing out an error.

CIVIL SURVEYING SAMPLE EXAMS
FOR THE CALIFORNIA SPECIAL CIVIL ENGINEER EXAMINATION
Second Edition

Current printing of this edition: 1

Printing History

edition number	printing number	update
1	1	New book.
2	1	New edition.

Printed in the United States of America

PPI
1250 Fifth Avenue, Belmont, CA 94002
(650) 593-9119
www.ppi2pass.com

Library of Congress Cataloging-in-Publication Data
Boniface, Peter R., 1937-
Civil surveying sample exams for the California special civil engineer examination / Peter R. Boniface.--2nd ed.
p. cm.
ISBN-13: 978-1-59126-100-1
ISBN-10: 1-59126-100-7
1. Surveying--California--Examinations, questions, etc. 2. Surveyors--Certification--California--Study guides. I. Monroe, James R., 1959-. II. Title.

TA537.B66 2007
526.9076--dc22

2006050999

Table of Contents

Preface and Acknowledgments

Civil Surveying Sample Exams is designed to simulate the exam experience and give you the authentic preparation you need for the California Special Civil Engineer Surveying Examination. The problems in this book are based on the Engineering Surveying Test Plan as published on the website of the California Board for Professional Engineers and Land Surveyors. The exam syllabus is extremely broad and touches on topics that most civil engineers have not studied during their undergraduate education. This is why I've provided you with two sample exams—so you can double your problem-solving practice.

The 50 problems in each sample exam mirror the actual exam problems in subject matter, length, and degree of difficulty. Although past examinations are not generally made publicly available, the published test plan does list all topics that appear on the exam and gives the following percentages for the coverage of groups of topics: calculations, 33%; surveying office procedures, 41%; equipment and field activities, 15%; and field measurement, 11%. I have attempted to match these percentages in the sample exams contained here.

I would like to thank Michael Lee, PE, PLS, and George M. Cole, PE, PLS, for performing the technical reviews of the first and second editions of this book, respectively. I would also like to thank the PPI editorial and production staffs for their hard work and patience in helping to put this second edition together. Finally, I would like to thank James R. Monroe, Jr., PE, for writing *Civil Surveying Practice Exams*, from which this book is derived.

Peter R. Boniface, PhD, PLS

Introduction

This book is designed to prepare you for the California Special Civil Engineer Examination for Engineering Surveying. California is currently the only state that requires candidates for licensure as civil engineers to pass such an exam in addition to other registration requirements.

You should begin your exam preparation by reviewing a wide range of textbooks in this field. You'll need a solid understanding of the fundamentals and principles of engineering surveying. As others have found, excellent results come from studying *Surveying Principles for Civil Engineers*, by Paul A. Cuomo, PLS; *1001 Solved Surveying Fundamentals Problems*, by Jan Van Sickle, PLS; and *120 Solved Surveying Problems for the California Special Civil Engineer Examination*, by Peter R. Boniface, PhD, PLS; these books are published by PPI. This book is designed to complement the other publications.

Civil Surveying Sample Exams acquaints you with the test plan adopted by the California Board for Professional Engineers and Land Surveyors. The intent of these practice exams is to measure your preparedness. With them, you can appraise your knowledge and skills before taking the official exam. Solutions are presented with explanations of the relevant key points and essential steps for solving problems.

After taking the sample exams, you will be better able to identify your strengths and weaknesses in all subject areas of the Engineering Surveying Test Plan, which will help you make an informed decision regarding further review and preparation for the California Special Civil Engineer Examination for Engineering Surveying. When you concentrate and work on your weak areas, you will be far better prepared and, as a result, will improve your performance on the official exam.

Engineers who have taken the civil surveying exam frequently comment that there is not enough time to complete the questions. It is therefore imperative that you improve your problem-solving speed by becoming proficient in the use of a calculator—particularly with respect to computations involving angles. Angles, bearings, and distances are fundamental to virtually all survey applications. Unfortunately, most calculators do not easily handle degrees, minutes, and seconds, and the default angular unit is decimal degrees. You should make a point of learning the functions that enable rapid computation of trigonometric functions as well as addition, subtraction, and multiplication of angles directly in degrees, minutes, and seconds.

Pay special attention to the following topics, which are fundamental to surveying and form a necessary core of knowledge: coordinate geometry (COGO) functions (such as inverse, side shot, bearing-bearing intersection, traverse, areas), differential leveling, horizontal and vertical curves, earthwork volumes, and datums.

The learning style and approach of each examinee to understanding the exam subject matter is unique. For exam preparation, you may do self-study or take formal instruction, whichever suits you best. In addition, taking these sample exams in a simulated examination situation with constraints similar to those of the official exam can measure your level of readiness. There is no better way to increase your proficiency in and knowledge of all surveying subject areas covered in the exam and to improve your chances of passing than to test yourself using sample exams and practice problems.

The Nature of the Exam

The California Special Civil Engineer Examination is given twice a year, in April and October. It tests the entry-level competency of a candidate to practice civil engineering within the profession's acceptable standards for public safety. This exam is open book and is administered over a two and a half hour period. The exam contains 50 multiple-choice problems derived from content areas as outlined in the board-adopted Engineering Surveying Test Plan. For each question, you will be asked to select the best answer from four choices. You'll record your answers on a machine-scorable answer sheet that will be provided to you when the exam is administered. Your calculations should be performed in the official test book, and they will not be credited or scored. Also, answers marked in the official test book will not be scored, and additional time will not be permitted to transfer answers to the official machine-scorable answer sheet.

The point values of each exam question are printed in the official test booklet. Points are assigned depending on the significance, difficulty, and complexity of the question.

What to Take to the Exam

You should come to the exam prepared with the following materials. Be sure to set them out ahead of time and devise a convenient way to carry them into the examination room with you.

- admission notice
- photo ID
- reference materials (only as much as you can carry in one trip)
- NCEES-approved calculator
- spare calculator batteries
- scales
- triangles
- compass
- protractor

Test-Taking Strategy

The California Special Civil Engineer Examination for Engineering Surveying is a difficult test and requires thorough preparation in all areas, including multiple-choice test-taking techniques. Easy and difficult questions, with variable point values, are distributed throughout the exam. Besides contending with the nature and difficulty of the exam itself, many examinees spend too much time on difficult problems and leave insufficient time to answer the easy ones. You should avoid this. The following system can be very beneficial to you.

step 1: Work on the easy questions immediately and record your answers on your official machine-scorable answer sheet.

step 2: Work on questions that require minimal calculations and record your answers on your official machine-scorable answer sheet.

step 3: When you get to a question that looks "impossible" to answer, mark a "?" next to it in your official test booklet, and go ahead and guess. Mark your "guess" answer on your answer sheet and continue.

step 4: When you face a question that seems difficult but solvable, mark an "X" next to it in your official test booklet; you may need considerable time to search for relevant information in your books, references, or notes. Continue to the next question.

step 5: When you come to a question that is solvable but you know requires lengthy calculations or is time consuming, mark a "+" sign next to the question in your official test booklet. For this question, you know exactly where to look for relevant information in your books, references, or notes.

Based on the number of exam questions and allotted time, on the average you should not spend more than 2.5 minutes per question. Thus, a lengthy or "time-consuming" question is one that will take you more than 2 minutes and 30 seconds to answer. Quickly and confidently, decide whether a question should receive a "+" or an "X;" the intent of this test-taking strategy is to save you precious time.

After you have gone over the entire exam, your official test booklet will clearly show which questions you have already answered and those that still require your attention.

You should now go over the exam a second time, with the following approach in mind.

step 1: The best and most successful approach is to go back and tackle: (a) the "+" questions, (b) the "X" questions, and (c) the "?" questions. As you proceed, eliminate your pluses, Xs, and question marks.

step 2: Recheck your work for careless mistakes.

step 3: Set aside the last few minutes of your exam period to fill in a guess for any unanswered lines on your official machine-scorable answer sheet. There is no penalty for guessing. Only questions answered correctly will be counted toward your score.

How to Use This Book

Civil Surveying Sample Exams for the California Special Civil Engineer Examination should be used in conjunction with surveying textbooks. For optimal results, you should take these sample exams only after you have reviewed a wide range of textbooks on the surveying field and grasped the subject matter in depth. You should also become familiar with the content and topics of the exam, as outlined in the board-adopted test plan, and focus your study efforts on those areas. Finally, it is suggested that you solve as many sample problems as possible and consider applying some multiple-choice test-taking strategies as you take the sample exams.

You will benefit by simulating the official exam constraints and conditions and giving yourself two and a half hours (try using a timer) to answer the problems in each sample exam. If friends or associates are also preparing for the exam, you may want to get together for a group exam simulation. When it comes to scoring and going over the provided solutions, a group discussion will help you to understand the subject matter more thoroughly.

When taking a sample exam, follow these steps.

step 1: Use the answer sheet provided to record your answers. Do not look at the solutions until after you have completed the sample exam.

step 2: Set a timer for two and a half hours, and begin working.

step 3: When time is up, stop working on the problems. Or, if you finish early, recheck your work during the remaining time. The rechecking habit could serve you well on the official exam.

step 4: Use the answer key provided to check your results and determine your score. This step will enable you to tell what areas you may need to review for a better grasp of the subject matter.

step 5: Appraise your performance. If your point total is more than 60% of the perfect score (>150 points out of 250 for these sample exams), it can be considered a passing score.

step 6: Study the provided solutions, and review your textbooks and references for those areas where you either answered incorrectly or guessed. If you answered some questions correctly but feel a need to understand their concepts more in depth, study those areas as well.

Exam Scoring

For the California Special Civil Engineer Examination for Engineering Surveying, the board-adopted test plan lists the content areas of the exam and their assigned scoring percentages. The percentages assigned to each content area are the approximate proportion of total test points; however, the test plan does not reveal the total test points in advance (it varies from exam to exam). This makes it difficult to anticipate the exact number of problems for each content area of the exam.

The official exam is graded against a "cut score"—a predetermined minimum passing score that varies from exam to exam. Historically, if you score above 60% of the total examination point value, you have a chance of passing.

The sample exams in this book each adopt a total score of 250 for their 50 multiple-choice problems. The point values for each problem are given next to the problem statements. (Point values appear on the actual exam as well.) In this book, the point values are also printed on the answer keys. A point total of 150 or more on these sample exams can be considered a passing score. Your probability of success on the actual exam will increase when you score higher on these sample exams.

On the official exam, after initial scoring, any problem that the board finds to be flawed may be deleted. In the event of deletion, the point value of the deleted problem becomes zero and the total number of points possible on that exam is adjusted accordingly.

You will face the official exam with a higher probability of success by going through the scoring process on your practice exam. The scoring process will give you an idea of how to overcome your weaknesses and pass the exam.

Nomenclature

A	area	$\text{in}^2, \text{ft}^2, \text{yd}^2$
A'	area	yd^2
b	base	ft
BS	backsight	ft
C	correction factor	ft
C	cost	$
C_f	earth curvature correction	ft
CE	cost of excavation	$
CO	cost of overhaul	$
D	degree of curvature	deg
D	depth	ft
D	distance	ft
dep	departure	ft
dx	departure difference	ft
dy	latitude difference	ft
elev	elevation	ft
FS	foresight	ft
g	grade	%
h	height	ft
HD	horizontal distance	ft
HI	height of instrument	ft
I	deflection angle	deg
IFS	intermediate foresight	ft
k	constant	deg
L	length	ft
L	length of curve	ft
lat	latitude	ft
P_l	tension (pull) on tape	lbf
r	rate of change of grade	%/ft
R	radius	ft
R	radius of the earth	ft
R	range	ft
R_f	refraction correction	ft
S	section area	ft^2
SD	slope distance	ft
T	standard temperature of tape	°F
T	tangent distance	ft
T_l	temperature of tape	°F
V	volume	yd^3
w	unit weight of tape	lbf/ft
W	width	ft
x	distance	ft
x	tangent offset	ft
y	elevation	ft
y'	tangent offset	ft

Symbols

α	angle	deg
α	azimuth	deg
α	coefficient of thermal expansion of steel	1/°F
β	angle	deg
γ	angle	deg
θ	vertical angle	deg
σ	standard deviation	–

Subscripts

c	correction
BM	benchmark
BVC	beginning of vertical curve
l	length
m	mean or midway
PVI	point of vertical intersection
s	sag
S	section
TP	test point

Sample Exam 1
Problems

Answer Sheet

1. (A) (B) (C) (D)
2. (A) (B) (C) (D)
3. (A) (B) (C) (D)
4. (A) (B) (C) (D)
5. (A) (B) (C) (D)
6. (A) (B) (C) (D)
7. (A) (B) (C) (D)
8. (A) (B) (C) (D)
9. (A) (B) (C) (D)
10. (A) (B) (C) (D)
11. (A) (B) (C) (D)
12. (A) (B) (C) (D)
13. (A) (B) (C) (D)
14. (A) (B) (C) (D)
15. (A) (B) (C) (D)
16. (A) (B) (C) (D)
17. (A) (B) (C) (D)
18. (A) (B) (C) (D)
19. (A) (B) (C) (D)
20. (A) (B) (C) (D)
21. (A) (B) (C) (D)
22. (A) (B) (C) (D)
23. (A) (B) (C) (D)
24. (A) (B) (C) (D)
25. (A) (B) (C) (D)
26. (A) (B) (C) (D)
27. (A) (B) (C) (D)
28. (A) (B) (C) (D)
29. (A) (B) (C) (D)
30. (A) (B) (C) (D)
31. (A) (B) (C) (D)
32. (A) (B) (C) (D)
33. (A) (B) (C) (D)
34. (A) (B) (C) (D)
35. (A) (B) (C) (D)
36. (A) (B) (C) (D)
37. (A) (B) (C) (D)
38. (A) (B) (C) (D)
39. (A) (B) (C) (D)
40. (A) (B) (C) (D)
41. (A) (B) (C) (D)
42. (A) (B) (C) (D)
43. (A) (B) (C) (D)
44. (A) (B) (C) (D)
45. (A) (B) (C) (D)
46. (A) (B) (C) (D)
47. (A) (B) (C) (D)
48. (A) (B) (C) (D)
49. (A) (B) (C) (D)
50. (A) (B) (C) (D)

1 *(6 points)*

What is the area of a sector with a central angle of 47° within a curve that has a radius of 1100.00 ft?

(A) 10.3 ac
(B) 10.8 ac
(C) 11.0 ac
(D) 11.4 ac

2 *(4 points)*

Which of the following professional services can a civil engineer legally solicit in the state of California?

(A) boundary surveys
(B) construction staking
(C) infrastructure design
(D) all of the above

Refer to the following illustration for Probs. 3 through 5.

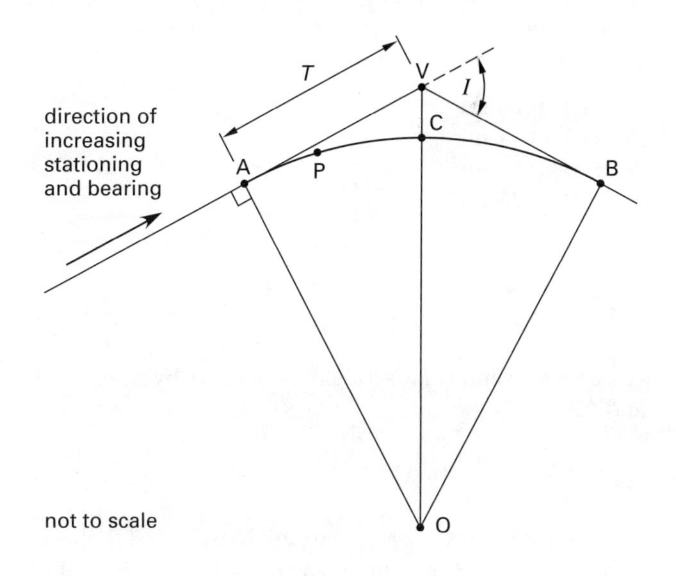

3 *(4 points)*

If the tangent distance, T, is 2658.55 ft and the deflection angle, I, is 56°, what is most nearly the radius of the curve?

(A) 1790 ft
(B) 3010 ft
(C) 5000 ft
(D) 5660 ft

4 *(4 points)*

If the degree of curvature (arc definition) is 1.1459°, what is most nearly the radius?

(A) 3200 ft
(B) 3800 ft
(C) 4200 ft
(D) 5000 ft

5 *(6 points)*

Point P lies on a 4000 ft radius curve. Point Q lies on the tangent to the curve, and line PQ is perpendicular to the tangent. The distance AQ is 100 ft. What is most nearly the distance PQ?

(A) 1.3 ft
(B) 2.5 ft
(C) 3.0 ft
(D) 9.5 ft

6 *(4 points)*

An aerial photograph was taken using a camera with a focal length of 6 in. The plane was flying at an altitude of 4000 ft above mean sea level, and the mean ground elevation was 400 ft above mean sea level. What is the scale of the photograph?

(A) 1 in:10 ft
(B) 1 in:50 ft
(C) 1 in:600 ft
(D) 1 ft:10 ft

7 *(4 points)*

In a boundary survey involving the reestablishment of corner monuments, if the survey measurements agree with those on an existing recorded map, the surveyor should prepare a

(A) corner record
(B) record of survey
(C) tract map
(D) parcel map

8 *(4 points)*

A line of magnetic bearing of N 12°32′ E and magnetic declination of 5° E is observed today. What was its magnetic bearing 100 years ago if the magnetic declination at that time was 2° E?

(A) N 9°32′ E
(B) N 12°29′ E
(C) N 12°35′ E
(D) N 15°32′ E

Refer to the following illustration, based on the U.S. Public Lands System, for Probs. 9 and 10.

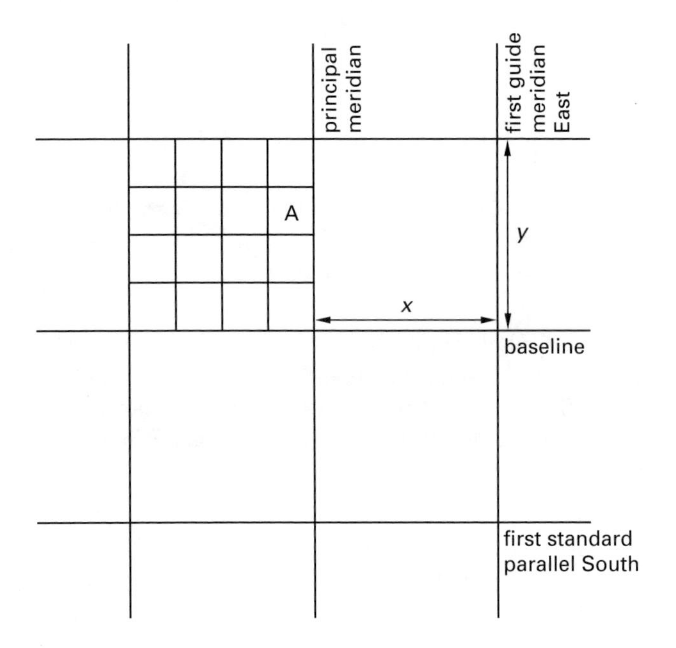

9 *(4 points)*

What designation is given to the section labeled A?

(A) township 3 North
(B) range 1 West
(C) township 3 North, range 1 West
(D) township 1 West, range 3 North

10 *(4 points)*

The term used for the area bounded by the labels x and y is a

(A) township
(B) range
(C) quadrangle
(D) section

11 *(4 points)*

If the overlap on a pair of aerial photos is 60% with a photo scale of 1 in:500 ft, the area covered by the two overlapping photos is most nearly

(A) 0.20 mi^2
(B) 0.44 mi^2
(C) 0.60 mi^2
(D) 0.75 mi^2

12 *(4 points)*

A highway centerline straight is defined by stations L and M.

$$\text{sta L} = 15{+}34.86 \text{ sta}$$
$$\text{sta M} = 17{+}09.08 \text{ sta}$$

A level is set up between stations L and M. The backsight to L is 3.78 ft. The foresight to M is 8.10 ft. The grade of line LM is most nearly

(A) 0.10%
(B) 2.5%
(C) 4.6%
(D) 6.8%

13 *(4 points)*

A data recorder is

(A) a handheld calculator that is used for manually entering field data
(B) a voice-activated device attached to a GPS receiver
(C) an electronic unit that automatically records data from a total station
(D) a CCD array that automatically records readings from a bar-coded leveling rod

Refer to the following illustration for Probs. 14 through 16.

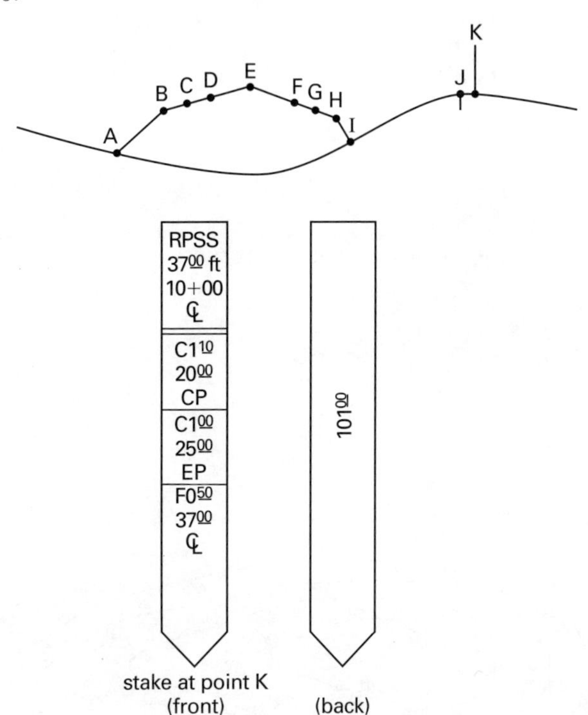

14 *(4 points)*

The first entry under the double strike line on the coded stake at point K indicates that the

(A) excavation required at point H is 1.1 ft
(B) finished grade elevation at point H is designed to be 1.1 ft below the elevation of the stake set at point J
(C) embankment required at point H is 1.1 ft
(D) difference in elevation between point I and point J is 1.1 ft

15 *(4 points)*

The second entry under the double strike line on the coded stake at point K indicates that the

(A) excavation at point F is 1.0 ft
(B) embankment at point F is 1.0 ft
(C) difference in elevation between point G and point F is 1.0 ft
(D) difference in elevation between the finished grade at point F is 1.0 ft below the elevation of the stake set at point J

16 *(4 points)*

The code on the back of the stake at point K indicates that the

(A) elevation of the top of the stake at point J is 101.00 ft
(B) offset distance from point K to point E is 101.00 ft
(C) design finished grade elevation of point E is 101.00 ft
(D) offset distance from point A to point I is 101.00 ft

17 *(4 points)*

In the state of California, only a licensed land surveyor is permitted to

(A) establish, in the field, the centerline of a proposed highway
(B) calculate the deflection angles required in establishing a traverse used as control in the design and construction of a proposed building
(C) set property corners in a new residential subdivision
(D) both (A) and (C)

18 *(8 points)*

A large parcel of land is to be subdivided such that the area of parcel 1 is 6.00 ac, as shown.

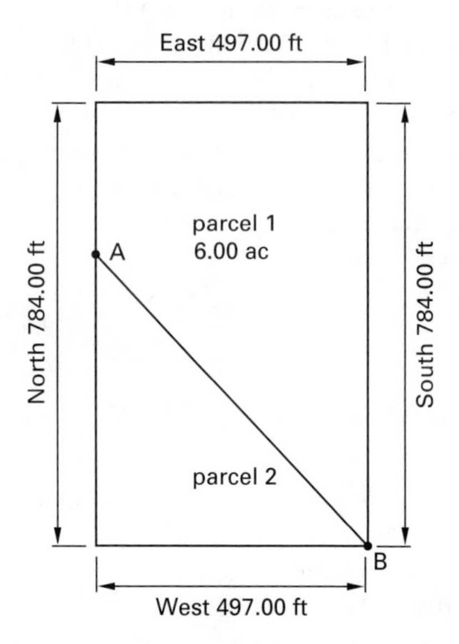

The length of boundary AB is most nearly

(A) 690 ft
(B) 720 ft
(C) 740 ft
(D) 750 ft

Refer to the following illustration for Probs. 19 through 21.

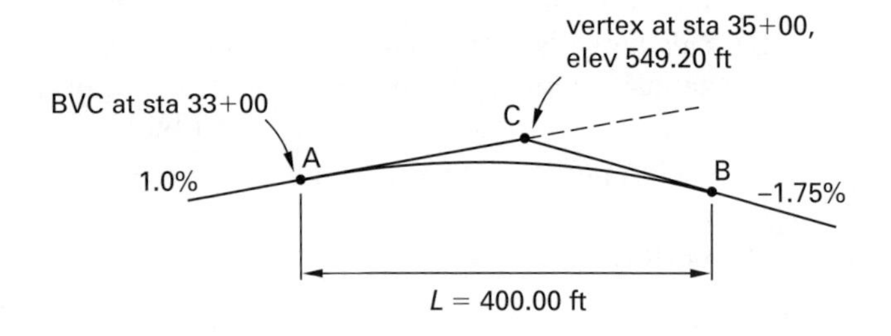

19 *(6 points)*

At what station does the curve crest?

(A) sta 33+93.71
(B) sta 34+35.40
(C) sta 34+45.45
(D) sta 35+11.00

20 *(8 points)*

What is most nearly the elevation difference between BVC (beginning of vertical curve) and a point on the curve at sta 35+00?

(A) 0.63 ft
(B) 0.97 ft
(C) 1.2 ft
(D) 33 ft

21 *(6 points)*

What is the highest elevation on the curve?

(A) 547.93 ft
(B) 548.44 ft
(C) 548.62 ft
(D) 549.38 ft

22 *(6 points)*

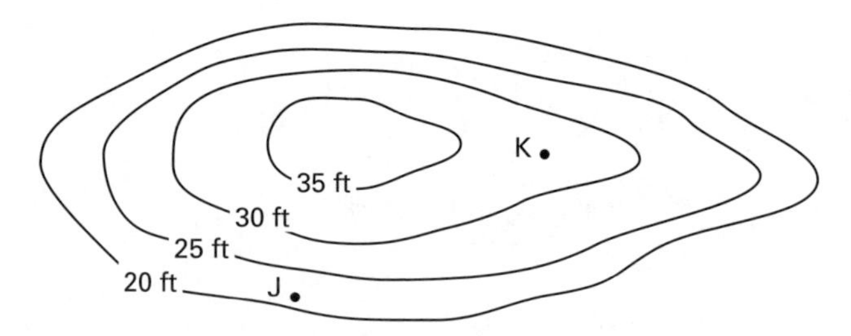

In the given topography, horizontal distance JK is 198 ft. The percent grade of line JK is most nearly

(A) 2%
(B) 4%
(C) 6%
(D) 8%

23 *(8 points)*

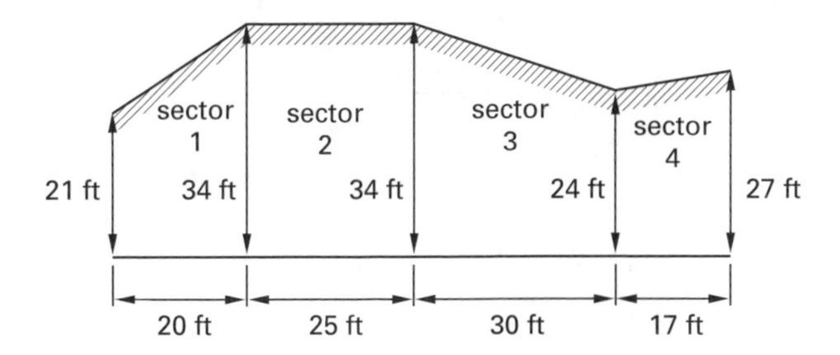

The area of the cross section shown is most nearly

(A) 100 yd^2
(B) 160 yd^2
(C) 240 yd^2
(D) 300 yd^2

24 *(4 points)*

When performing a leveling circuit, what is the purpose of positioning the instrument such that the foresight and backsight are approximately equal in length?

(A) Errors due to the line of sight not being perfectly horizontal tend to cancel each other.
(B) Errors due to the line of sight not being perfectly vertical tend to cancel each other.
(C) Errors due to the instrument's varying degree of focus when reading the rod tend to cancel each other.
(D) both (B) and (C)

25 *(6 points)*

A 200 ft horizontal tape hangs freely in a catenary shape. It is supported at the 80 ft mark in addition to its ends. The tension on the tape is 20 lbf, and the weight of the entire tape is 3 lbf 5 oz. The measured distance is 187.01 ft. What is the true distance?

(A) 186.06 ft
(B) 186.82 ft
(C) 186.96 ft
(D) 187.20 ft

26 *(6 points)*

The observed interior angles of a five-course traverse are

$$52°16'40''$$
$$89°06'20''$$
$$139°24'40''$$
$$41°37'00''$$
$$217°36'40''$$

If the traverse angles are balanced, what is the value of the last angle?

(A) 217°35′20″
(B) 217°36′24″
(C) 217°36′56″
(D) 217°38′00″

27 *(8 points)*

The following data pertains to an open four-course traverse.

side	distance	azimuth	point	coordinate y	coordinate x
			A	560.00	770.00
AB	394.59	81°14′54″			
			B		
BC	292.75	172°08′50″			
			C		
CD	332.87	212°44′26″			
			D		
DE	323.88	351°07′42″			
			E		

The (y, x) coordinates of point E are

(A) (370.05, 970.02)
(B) (687.54, 644.10)
(C) (749.95, 569.98)
(D) (760.02, 580.05)

28 *(6 points)*

A straight line 8.6 km in length is trig-leveled. What is most nearly the combined effect of earth curvature and refraction on this line?

(A) 16.40 ft
(B) 19.02 ft
(C) 20.07 ft
(D) 21.65 ft

29 *(6 points)*

For a curve, the following parameters apply.

intersection angle	14°10′00″
curve radius	2400 ft
PI (point of intersection) stationing	sta 24+15.9

Points are placed on the curve at every 100 ft station. The deflection angle to the first point on the curve (after the tangent-to-curve point, TC) is

(A) 0°59′30″
(B) 0°59′00″
(C) 0°58′58″
(D) 0°57′57″

30 *(6 points)*

For the region shown, the following parameters apply.

(y, x) coordinate of A	$(1675.24, 2546.09)$
(y, x) coordinate of B	$(1294.39, 2870.06)$
azimuth AP	95°34′20″
azimuth BP	48°00′00″

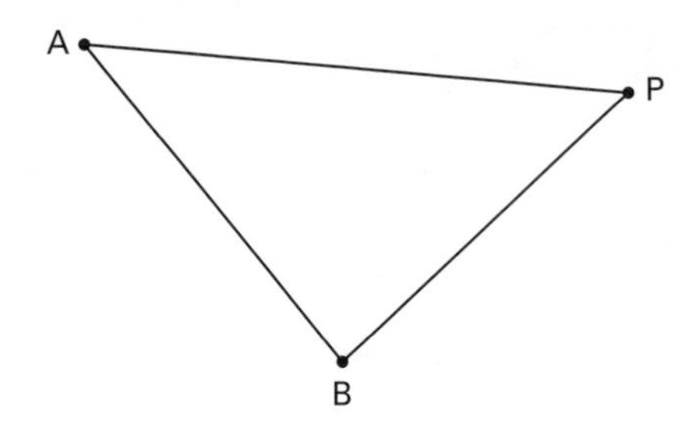

The (y, x) coordinates of point P are

(A) $(-1609.49, -3220.02)$
(B) $(1474.36, 3892.11)$
(C) $(1598.34, 3564.19)$
(D) $(1609.49, 3220.02)$

31 *(6 points)*

The (x_1, y_1) coordinates of a point 1 are $(410, 190)$ and the (x_2, y_2) coordinates of a point 2 are $(87, 290)$. The azimuth 1-2 is

(A) −72°47′51″
(B) −17°12′09″
(C) 287°12′09″
(D) 342°47′51″

32 *(6 points)*

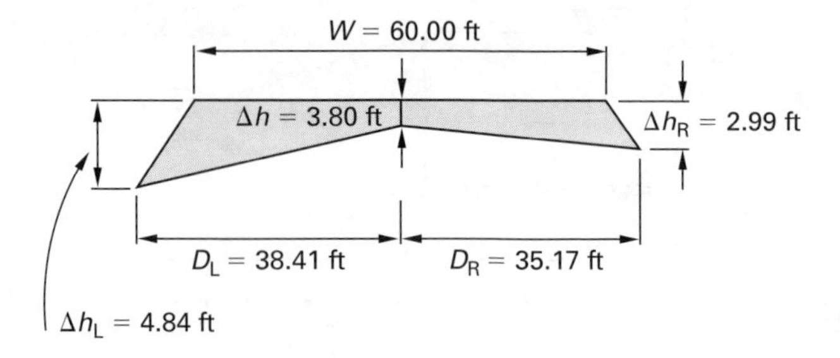

The area of the given roadway cross section is most nearly

(A) 190 ft^2
(B) 260 ft^2
(C) 300 ft^2
(D) 370 ft^2

33 *(4 points)*

A soccer field 120 yd in length measures 9.144 mm on an aerial photo. The scale of the aerial photo is most nearly

(A) 1:4000
(B) 1:10,000
(C) 1:12,000
(D) 1:110,000

34 *(4 points)*

Before computing a traverse on NAD 83, measured distances must be corrected for which of the following?

(A) grid scale factor
(B) sea-level scale factor
(C) neither (A) nor (B)
(D) both (A) and (B)

35 *(4 points)*

The most common elevation datum for modern surveys is

(A) NAD 27
(B) NAD 83
(C) NGVD 29
(D) NGVD 88

36 *(4 points)*

A total station with a precision of 0.5 in±10 ppm is used to measure a line 1200 ft long. What is the precision of this measurement in inches?

(A) 0.5 in
(B) 0.6 in
(C) 1.9 in
(D) 2.0 in

37 *(4 points)*

A closed traverse totaling 3250 ft in length closes onto a point with known (x, y) coordinates (2459.16 ft, 5211.90 ft). The computed coordinates of this point after the computation of the last traverse course are (2459.38 ft, 5211.71 ft). The accuracy of the traverse is most nearly

(A) 1:8000
(B) 1:11,000
(C) 1:12,000
(D) 1:15,000

38 *(6 points)*

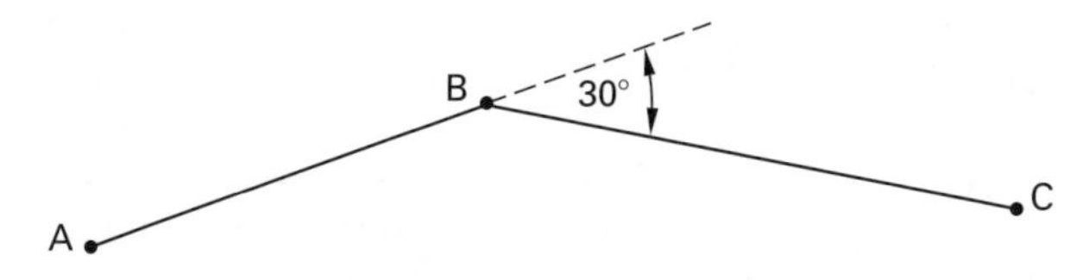

For the plan view shown, distance AB was measured as 145.76 ft with a slope of 2°20′, and distance BC was measured as 189.44 ft with a slope of 3°10′. The horizontal distance AC is

(A) 96.30 ft
(B) 309.57 ft
(C) 323.58 ft
(D) 334.79 ft

39 *(6 points)*

A line AB was measured three times as follows.

1st measurement: 597.410 ft at a temperature of 83°F
2nd measurement: 597.320 ft at a temperature of 98°F
3rd measurement: 597.500 ft at a temperature of 49°F

The standard temperature of the tape is 68°F, and the coefficient of expansion of steel is 6.45×10^{-6}°F^{-1}. The most probable length of line AB is most nearly

(A) 597.33 ft
(B) 597.38 ft
(C) 597.44 ft
(D) 597.49 ft

40 *(4 points)*

A line measures 2189.01 ft at an elevation of 4100 ft above sea level. It is to be used in a state-plane project. The radius of the earth is 20,906,000 ft. The length of the line at sea level is

(A) 2183.79 ft
(B) 2188.58 ft
(C) 2189.44 ft
(D) 2194.23 ft

41 *(4 points)*

For the computation of an earthwork quantity between two cross sections using the prismoidal formula, the middle section dimensions are obtained by

(A) measurement in the field
(B) averaging the dimensions of the two end cross sections
(C) averaging the dimensions of the two end cross sections using a weighted average
(D) dimensions are not required (the areas of the two end cross sections need only be averaged and used in the formula)

42 *(8 points)*

Dirt is excavated from a borrow pit approximately 80 ft by 60 ft by 5 ft deep and moved 800 ft away. The cost of excavation is \$2.00/yd^3. The cost of overhaul for any amount over five stations is \$0.40/yd^3 per station. The first five stations are free. What is most nearly the total cost of the move?

(A) \$2800
(B) \$3000
(C) \$3500
(D) \$4600

43 *(4 points)*

The minimum ground control required to map from an overlapping pair of aerial photos is

(A) both a plan point and a height point in each corner of the overlap
(B) three plan and three height points, not in a straight line, on the overlap
(C) three height points, not in a straight line, and two plan points on the overlap
(D) two plan points in opposite corners of the overlap and two height points, also in opposite corners of the overlap

44 *(4 points)*

When the owners of adjacent properties wish to permanently alter a common boundary, the required survey process is known as

(A) lot-line adjustment
(B) legal consolidation
(C) boundary realignment
(D) boundary agreement

45 *(4 points)*

The current standard boundary monument in California is a license tag attached to the top of a

(A) 2 in by 2 in redwood stake
(B) 1 in iron pipe, set flush with the soil
(C) 2 ft rebar
(D) spike, set in concrete

46 *(4 points)*

The measured distance between two points A and B is 3864.7 ft. The combined earth curvature and refraction correction along this line is

(A) −0.41 ft
(B) −0.31 ft
(C) 0.31 ft
(D) 0.41 ft

47 *(4 points)*

A 12 ft leveling rod is held next to a roof such that the top of the rod (12.00 ft mark) is level with the edge of the roof. A level is set up and a sight is taken to the rod, which reads 4.89 ft. The instrument is turned and a temporary benchmark is sighted; the rod reading is 7.90 ft. The height difference between the benchmark and the edge of the roof is

(A) 0.79 ft
(B) 1.79 ft
(C) 14.01 ft
(D) 15.01 ft

48 *(4 points)*

The rectangular method of defining the location of a 6 by 6 group of sections in the Public Lands System is known as

(A) northing and easting
(B) township and range
(C) north section and east section
(D) latitude and longitude

49 *(4 points)*

A tract of land that is described in a legal description by "metes and bounds" is said to be described by

(A) properties that bound or are adjacent to the property in question
(B) state-plane coordinates
(C) corner points established by reliable witnesses
(D) bearings and lengths of the boundaries

50 *(6 points)*

A distance is measured six times with the following results.

$x_1 = 178.45$ ft $x_4 = 178.44$ ft
$x_2 = 178.48$ ft $x_5 = 178.74$ ft
$x_3 = 178.49$ ft $x_6 = 178.47$ ft

The standard error of a single measurement is most nearly

(A) 0.001 ft
(B) 0.005 ft
(C) 0.007 ft
(D) 0.02 ft

Sample Exam 1
Solutions

Answer Key

No.					Point Value
1.	A	B	C	●	6
2.	A	B	C	●	4
3.	A	B	●	D	4
4.	A	B	C	●	4
5.	●	B	C	D	6
6.	A	B	●	D	4
7.	●	B	C	D	4
8.	A	B	C	●	4
9.	A	B	●	D	4
10.	A	B	●	D	4
11.	A	●	C	D	4
12.	A	●	C	D	4
13.	A	B	●	D	4
14.	A	●	C	D	4
15.	A	B	C	●	4
16.	●	B	C	D	4
17.	A	B	●	D	4
18.	A	●	C	D	8
19.	A	B	●	D	6
20.	●	B	C	D	6
21.	●	B	C	D	8
22.	A	B	●	D	8
23.	A	B	C	●	6
24.	●	B	C	D	4
25.	A	B	●	D	6

No.					Point Value
26.	A	●	C	D	6
27.	●	B	C	D	8
28.	●	B	C	D	6
29.	A	B	●	D	6
30.	A	B	C	●	6
31.	A	B	●	D	6
32.	A	●	C	D	6
33.	A	B	●	D	4
34.	A	B	C	●	4
35.	A	B	C	●	4
36.	A	●	C	D	4
37.	A	●	C	D	4
38.	A	B	●	D	6
39.	A	B	●	D	6
40.	A	●	C	D	4
41.	A	●	C	D	4
42.	●	B	C	D	8
43.	A	B	●	D	4
44.	●	B	C	D	4
45.	A	●	C	D	4
46.	A	B	●	D	4
47.	A	B	C	●	4
48.	A	●	C	D	4
49.	A	B	C	●	4
50.	A	B	C	●	6

Perfect Score = 250 points
Point Total: ______________

Note: A point total of greater than 150 can be considered a passing score on this sample exam.

1 *The answer is (D).*

$$\frac{A}{\pi R^2} = \frac{I}{360°}$$

Rearranging to solve for A,

$$A = \left(\frac{I}{360°}\right)\pi R^2 = \left(\frac{47°}{360°}\right)\pi(1100 \text{ ft})^2$$
$$= 496{,}284 \text{ ft}^2$$

Convert this area to acres.

$$A = \frac{496{,}284 \text{ ft}^2}{43{,}560 \ \frac{\text{ft}^2}{\text{ac}}} = 11.39 \text{ ac}$$

2 *The answer is (D).*

A licensed civil engineer can legally solicit work that his license does not cover, provided the work is ultimately performed under the supervision of a person with the appropriate license. (The codes are B&P 6735 and 6738.)

3 *The answer is (C).*

$$T = R\tan\frac{I}{2}$$
$$2658.55 \text{ ft} = R\tan\frac{56°}{2}$$

Rearranging to solve for R,

$$R = \frac{2658.55 \text{ ft}}{\tan\frac{56°}{2}}$$
$$= 5000.01 \text{ ft} \quad (5000 \text{ ft})$$

4 *The answer is (D).*

Using the conversion equation expressing degree of curvature, D, as a function of radius, R,

$$D = \frac{(180°)(100 \text{ ft})}{\pi R}$$

Rearranging to solve for R,

$$R = \frac{(180°)(100 \text{ ft})}{\pi D} = \frac{(180°)(100 \text{ ft})}{\pi(1.1459°)}$$
$$= 5000 \text{ ft}$$

5 *The answer is (A).*

Using tangent offsets,

$$x = R\sin\alpha$$
$$100 \text{ ft} = (4000 \text{ ft})\sin\alpha$$

Rearranging,

$$\sin\alpha = \frac{100 \text{ ft}}{4000 \text{ ft}} = \frac{1}{40}$$

Solving for α,

$$\alpha = \arcsin\frac{1}{40} = 1.433°$$
$$y = R(1 - \cos\alpha) = (4000 \text{ ft})(1 - \cos 1.433°)$$
$$= 1.25 \text{ ft} \quad (1.3 \text{ ft})$$

6 *The answer is (C).*

$$\text{scale} = \frac{\text{focal length}}{\text{altitude} - \text{ground elevation}}$$
$$= \frac{6 \text{ in}}{4000 \text{ ft} - 400 \text{ ft}}$$
$$= \frac{1 \text{ in}}{600 \text{ ft}} \quad (1 \text{ in:600 ft})$$

7 *The answer is (A).*

In order to simplify the preparation of maps where there is no discrepancy in the surveyed data, the surveyor can prepare a corner record, which can be hand-drawn on a standard form.

8 *The answer is (D).*

The true bearing of the line today is

$$\text{N } 12°32' \text{ E} + 05°00' \text{ E} = \text{N } 17°32' \text{ E}$$

The magnetic bearing of the line 100 years ago was

$$\text{N } 17°32' \text{ E} - 02°00' \text{ E} = \text{N } 15°32' \text{ E}$$

9 *The answer is (C).*

The U.S. Public Lands System would define the area as township 3 North, range 1 West.

10 *The answer is (C).*

The area bounded by the labels x and y is called a quadrangle.

11 *The answer is (B).*

A standard aerial photo is 9 in by 9 in. From the photo scale, 1 in on the photo represents 500 ft on the ground. Therefore, 9 in on the photo equals (9 in)(500 ft/in), or 4500 ft, on the ground.

The length of a side is 60% of the 9 in distance. This can be calculated as

$$L_{60\%} = (0.6)(4500 \text{ ft}) = 2700 \text{ ft}$$

The area covered by the overlapping photos is

$$A = \frac{(4500 \text{ ft})(2700 \text{ ft})}{\left(5280 \, \frac{\text{ft}}{\text{mi}}\right)^2} = 0.44 \text{ mi}^2$$

12 *The answer is (B).*

$$\begin{aligned} D_{\text{LM}} &= (\text{sta M}) - (\text{sta L}) \\ &= (17{+}09.08 \text{ sta}) - (15{+}34.86 \text{ sta}) \\ &= 174.22 \text{ ft} \\ \Delta h_{\text{LM}} &= \text{FS} - \text{BS} \\ &= 8.10 \text{ ft} - 3.78 \text{ ft} \\ &= 4.32 \text{ ft} \\ g_{\text{LM}} &= \frac{\Delta h_{\text{LM}}}{D_{\text{LM}}} \times 100\% \\ &= \frac{4.32 \text{ ft}}{174.22 \text{ ft}} \times 100\% \\ &= 2.48\% \quad (2.5\%) \end{aligned}$$

13 *The answer is (C).*

Calculators are sometimes used as data recorders, but data is never input into them manually. Global positioning system (GPS) receivers do not have voice-activated data recorders. Charged coupled device (CCD) arrays are used to automatically record readings on barcoded rods, but data recording refers to the automatic recording of survey data such as horizontal angles, distances, and vertical angles.

A data recorder is an integral part of a total station. Measured data such as slope distance, horizontal angles, and vertical angles as well as computed data such as vertical distances and horizontal distances are automatically stored in a data recorder for future downloading to a PC.

14 *The answer is (B).*

The first entry indicates that the finished grade elevation at point H is designed to be 1.1 ft below the elevation of the stake set at point J.

15 *The answer is (D).*

The second entry indicates that the difference in elevation between the finished grade at point F is 1.0 ft below the elevation of the stake set at point J.

16 *The answer is (A).*

The code indicates that the elevation of the top of the stake at point J is 101.00 ft.

17 *The answer is (C).*

Of the practices listed, in California, only the practice of setting property corners in a new residential subdivision is restricted to licensed land surveyors.

18 *The answer is (B).*

The area of the entire lot is

$$\begin{aligned} A_{\text{total}} = bh &= \frac{(784.00 \text{ ft})(497.00 \text{ ft})}{43{,}560 \, \frac{\text{ft}^2}{\text{ac}}} \\ &= 8.9451 \text{ ac} \end{aligned}$$

Since parcel 1 must contain 6 ac, the area of parcel 2 must be

$$\begin{aligned} A_2 &= A_{\text{total}} - A_1 \\ &= 8.9451 \text{ ac} - 6.00 \text{ ac} \\ &= 2.9451 \text{ ac} \end{aligned}$$

Convert this area to square feet.

$$(2.9451 \text{ ac})\left(43{,}560 \ \frac{\text{ft}^2}{\text{ac}}\right) = 128{,}288.56 \text{ ft}^2$$

Parcel 2 is a right triangle with area

$$\begin{aligned} A &= \frac{1}{2}bh \\ 128{,}288.56 \text{ ft}^2 &= \frac{1}{2}(497 \text{ ft})h \end{aligned}$$

Rearranging to solve for h,

$$\begin{aligned} h &= \frac{128{,}228.56 \text{ ft}^2}{\left(\frac{1}{2}\right)(497 \text{ ft})} \\ &= 516.25 \text{ ft} \end{aligned}$$

The hypotenuse $\overline{\text{AB}}$ can now be found.

$$\begin{aligned} (\overline{\text{AB}})^2 &= h^2 + b^2 \\ &= (516.25 \text{ ft})^2 + (497.00 \text{ ft})^2 \\ &= 513{,}523.06 \text{ ft}^2 \\ \overline{\text{AB}} &= \sqrt{513{,}523.06 \text{ ft}^2} \\ &= 716.61 \text{ ft} \quad (720 \text{ ft}) \end{aligned}$$

19 *The answer is (C).*

To find where the curve will crest,

$$\begin{aligned} x &= \frac{-g_1}{r} \\ g_1 &= 1.0\% \\ r &= \frac{g_2 - g_1}{L} = \frac{-1.75\% - 1.0\%}{4 \text{ sta}} \\ &= -0.6875\%/\text{sta} \\ x &= \frac{-1.0\%}{\frac{-0.6875\%}{\text{sta}}} \\ &= 1.4545 \text{ sta from BVC} \end{aligned}$$

The curve will crest at (sta 33+00) + (sta 1+45.45), or sta 34+45.45.

20 *The answer is (A).*

Begin by computing the equation of the curve.

$$\begin{aligned} r &= \frac{g_2 - g_1}{L} = \frac{-1.75\% - 1.0\%}{4 \text{ sta}} \\ &= -0.6875\%/\text{sta} \quad [\text{same as } -0.6875 \text{ ft/sta}^2] \\ y &= \left(\frac{r}{2}\right)x^2 + g_1 x \\ &= \left(\frac{\frac{-0.6875\%}{\text{sta}}}{2}\right)x^2 + (1.0\%)x \\ &= \left(-0.34375 \ \frac{\text{ft}}{\text{sta}^2}\right)(35 \text{ sta} - 33 \text{ sta})^2 \\ &\quad + \left(1.0 \ \frac{\text{ft}}{\text{sta}}\right)(35 \text{ sta} - 33 \text{ sta}) \\ &= 0.625 \text{ ft} \quad (0.63 \text{ ft}) \end{aligned}$$

21 *The answer is (A).*

Applying the equation for the curve,

$$y = \left(\frac{r}{2}\right)x^2 + g_1 x + \text{elev}_{\text{BVC}}$$

The curve crests at

$$\begin{aligned} x &= \frac{-g_1}{r} \\ r &= \frac{g_2 - g_1}{L} = \frac{-1.75\% - 1.0\%}{4 \text{ sta}} \\ &= -0.6875\%/\text{sta} \quad [\text{same as } -0.6875 \text{ ft/sta}^2] \\ x &= \frac{-1.0\%}{\frac{-0.6875\%}{\text{sta}}} \\ &= 1.4545 \text{ sta from B} \end{aligned}$$

So,

$$\begin{aligned} y &= \left(\frac{-0.6875 \ \frac{\text{ft}}{\text{sta}^2}}{2}\right)(1.4545 \text{ sta})^2 \\ &\quad + \left(1.0 \ \frac{\text{ft}}{\text{sta}}\right)(1.4545 \text{ sta}) + 547.20 \text{ ft} \\ &= 547.93 \text{ ft} \end{aligned}$$

22 *The answer is (C).*

K is midway between the 35 ft and 30 ft contours. Its elevation, y_K, is therefore 32.5 ft.

Similarly, linearly interpolating J between the 20 ft and 25 ft contours gives an elevation, y_J, of about 21 ft (graphical interpolation is approximate).

$$\begin{aligned}\Delta y_{JK} &= y_K - y_J = 32.5 \text{ ft} - 21.0 \text{ ft}\\ &= 11.5 \text{ ft}\end{aligned}$$

$$\begin{aligned}\% \text{ grade} &= \frac{\Delta y_{JK}}{x_{horiz}} \times 100\% = \frac{11.5 \text{ ft}}{198 \text{ ft}} \times 100\%\\ &= 5.81\% \quad (6\%)\end{aligned}$$

23 *The answer is (D).*

The area of sector 1 is

$$\begin{aligned}A &= \left(\frac{21 \text{ ft} + 34 \text{ ft}}{2}\right)(20 \text{ ft})\left(\frac{1 \text{ yd}}{3 \text{ ft}}\right)^2\\ &= 61.1 \text{ yd}^2\end{aligned}$$

The area of the remaining three sectors is

$$\begin{aligned}A' &= \left(\frac{34 \text{ ft} + 34 \text{ ft}}{2}\right)(25 \text{ ft}) + \left(\frac{34 \text{ ft} + 24 \text{ ft}}{2}\right)\\ &\quad \times (30 \text{ ft}) + \left(\frac{24 \text{ ft} + 27 \text{ ft}}{2}\right)(17 \text{ ft})\\ &= (2153.5 \text{ ft}^2)\left(\frac{1 \text{ yd}}{3 \text{ ft}}\right)^2\\ &= 239.3 \text{ yd}^2\end{aligned}$$

The area of the cross section is then

$$\begin{aligned}A_{\text{cross section}} &= A + A' = 61.1 \text{ yd}^2 + 239.3 \text{ yd}^2\\ &= 300.4 \text{ yd}^2 \quad (300 \text{ yd}^2)\end{aligned}$$

24 *The answer is (A).*

The instrument is positioned in that manner because errors due to the line of sight not being perfectly horizontal tend to cancel each other.

25 *The answer is (C).*

The true distance is shorter than the measured distance, due to sag in the tape. The correction for sag, C_s, is calculated using

$$C_s = \frac{w^2 L_s^3}{24 P_l^2}$$

Converting the portion of the weight of the tape that was measured in ounces,

$$w_{oz} = (5 \text{ oz})\left(\frac{1 \text{ lbf}}{16 \text{ oz}}\right) = 0.3125 \text{ lbf}$$

A common error is to use the weight of the entire tape instead of the weight of 1 ft of tape, w. Since w is the weight of a 1 ft length of tape,

$$\begin{aligned}w &= \frac{W}{L} = \frac{3 \text{ lbf} + 0.3125 \text{ lbf}}{200 \text{ ft}}\\ &= 0.016563 \text{ lbf/ft}\end{aligned}$$

There will be two sag corrections: one for an 80 ft bay (0–80 ft) and one for a 107 ft bay (80–187 ft). (It is important to count the correct number of bays (sags); for example, one support equals two bays.)

$$\begin{aligned}C_{s,\text{total}} &= C_{s,1} + C_{s,2}\\ &= \frac{\left(0.016563 \frac{\text{lbf}}{\text{ft}}\right)^2 (80 \text{ ft})^3}{(24)(20 \text{ lbf})^2}\\ &\quad + \frac{\left(0.016563 \frac{\text{lbf}}{\text{ft}}\right)^2 (107 \text{ ft})^3}{(24)(20 \text{ lbf})^2}\\ &= 0.05 \text{ ft}\end{aligned}$$

$$\begin{aligned}D_{\text{true}} &= D_{\text{measured}} - C_s\\ &= 187.01 \text{ ft} - 0.05 \text{ ft}\\ &= 186.96 \text{ ft}\end{aligned}$$

26 *The answer is (B).*

First, sum the observed angles.

$$\begin{aligned}\sum \begin{matrix}\text{observed}\\ \text{angles}\end{matrix} &= 52°16'40'' + 89°06'20'' + 139°24'40''\\ &\quad + 41°37'00'' + 217°36'40''\\ &= 540°01'20''\end{aligned}$$

If it were possible to observe a traverse with no errors, this sum would exactly equal $(n - 2)180°$, where n is the number of sides or courses. Here,

$$(n - 2)180° = (5 - 2)(180°) = 540°$$

However, due to the accumulation of observation errors, the sum will differ slightly from this value. This difference is the misclose. Therefore,

$$\text{misclose} = 540°00'00'' - 540°01'20'' = -1'20''$$

Divide the misclose by n (i.e., 5) to obtain the correction for each angle.

$$\text{correction} = \frac{\text{misclose}}{n} = \frac{-1'20''}{5} = -16''$$

Each angle is adjusted by this amount ($-16''$), giving the corrected or balanced angles. The balanced last angle will be $217°36'40'' - 16''$, or $217°36'24''$.

27 *The answer is (A).*

Compute the latitude difference, dy, and departure difference, dx.

$$dx = D\sin\alpha$$
$$dy = D\cos\alpha$$

Compute the coordinates of points B, C, D, and E by sequentially adding dy and dx, the latitude and departure differences, which are in italic in the table. For example,

$$dy_{\text{AB}} = 60.038$$
$$y_{\text{B}} = 560.000 + 60.038$$
$$dx_{\text{AB}} = 389.996$$
$$x_{\text{B}} = 770.000 + 389.996$$

				coordinate	
side	distance	azimuth	point	y	x
			A	560.00	770.00
AB	394.59	81°14′54″		*60.038*	*389.996*
			B	620.038	1159.996
BC	292.75	172°08′50″		*−290.005*	*39.998*
			C	330.033	1199.994
CD	332.87	212°44′26″		−279.986	−180.028
			D	50.047	1019.966
DE	323.88	351°07′42″		*320.005*	*−49.949*
			E	370.052	970.017
				(370.05)	(970.02)

28 *The answer is (A).*

The standard equations for distances given in feet are

$$C_f = 0.0239\left(\frac{D}{1000}\right)^2$$
$$R_f = -0.0033\left(\frac{D}{1000}\right)^2$$

The solution requires that the distance of the 8.6 km line be given in feet.

$$D_{\text{ft}} = (8.6\text{ km})\left(3280.8\ \frac{\text{ft}}{\text{km}}\right) = 28{,}215\text{ ft}$$

The earth curvature correction and refraction correction can then be determined. (The refraction component is negative and opposite in sign to the earth curvature correction.)

$$C_f = (0.0239)\left(\frac{28{,}215\text{ ft}}{1000}\right)^2 = 19.03\text{ ft}$$

$$R_f = (-0.0033)\left(\frac{28{,}215\text{ ft}}{1000}\right)^2 = -2.63\text{ ft}$$

$$\begin{aligned}\text{combined correction} &= C_f + R_f \\ &= 19.03\text{ ft} + (-2.63\text{ ft}) \\ &= 16.40\text{ ft}\end{aligned}$$

29 *The answer is (C).*

Compute the tangent distance and stationing of the TC point. The curve distance to the first point on the curve is then easily calculated by considering the next 100 ft point after the TC stationing.

$$\begin{aligned}T &= R\tan\frac{I}{2} \\ &= (2400\text{ ft})\tan 7°05'00'' \\ &= (2400\text{ ft})(0.124261) \\ &= 298.23\text{ ft}\end{aligned}$$

$$\begin{aligned}\text{sta TC} &= 24.1590\text{ sta} - 2.9823\text{ sta} \\ &= 21.1767\text{ sta}\end{aligned}$$

Therefore, the first point on the curve is sta 22+00.00. The distance along curve to first point is

$$\begin{aligned}T' &= 22.0000\text{ sta} - 21.1767\text{ sta} \\ &= 82.33\text{ ft}\end{aligned}$$

Compute the central angle, θ, from the curve distance to the first curve point. Then divide by 2 to get the intersection angle.

$$\begin{aligned}\theta &= \frac{T'}{R} = \frac{82.33\text{ ft}}{2400\text{ ft}} \\ &= (0.034304\text{ rad})\left(\frac{180°}{\pi\text{ rad}}\right) \\ &= 1.9655° \quad (1°57'56'')\end{aligned}$$

$$\alpha_{\text{TC}} = \frac{\theta}{2} = \frac{1.9655°}{2} = 0°58'58''$$

30 *The answer is (D).*

A bearing/bearing intersection is required to compute $(y_{\text{P}}, x_{\text{P}})$. The equations are

$$y_{\text{P}} = \frac{(x_{\text{A}} - x_{\text{B}}) - y_{\text{A}}\tan\alpha_{\text{AP}} + y_{\text{B}}\tan\alpha_{\text{BP}}}{\tan\alpha_{\text{BP}} - \tan\alpha_{\text{AP}}}$$

$$= \frac{(2546.09 - 2870.06) - 1675.24\tan 95°34'20'' + 1294.39\tan 48°00'00''}{\tan 48°00'00'' - \tan 95°34'20''}$$

$$= 1609.49$$

$$x_{\text{P}} = (y_{\text{P}} - y_{\text{A}})\tan\alpha_{\text{AP}} + x_{\text{A}} = (1609.49 - 1675.24)\tan 95°34'20'' + 2546.09 = 3220.02$$

31 *The answer is (C).*

The formula for the azimuth, α, is

$$\alpha = \tan^{-1}\frac{x_2 - x_1}{y_2 - y_1} + k$$

If $y_2 - y_1$ is positive, k will be 0° or 360°.

If $y_2 - y_1$ is negative, k will be 180°.

$$\alpha = \tan^{-1}\frac{87 - 410}{290 - 190} + k = \tan^{-1}(-3.23) + k$$

Since $y_2 - y_1$ is positive, k is 0° or 360°. But the inverse tangent is $-72°47'51''$. Therefore, k must equal 360°, since an azimuth must be positive.

$$\alpha = -72°47'51'' + 360° = 287°12'09''$$

32 *The answer is (B).*

The area of this "three-level" section is

$$A = \Delta h\left(\frac{D_{\text{L}} + D_{\text{R}}}{2}\right) + W\left(\frac{\Delta h_{\text{L}} + \Delta h_{\text{R}}}{4}\right)$$

$$= (3.80\text{ ft})\left(\frac{38.41\text{ ft} + 35.17\text{ ft}}{2}\right) + (60.00\text{ ft})\left(\frac{4.84\text{ ft} + 2.99\text{ ft}}{4}\right)$$

$$= 257.25\text{ ft}^2 \quad (260\text{ ft}^2)$$

33 *The answer is (C).*

First, convert the ground distance and photo distance to the same unit (inches).

$$D_{\text{ground}} = (120\text{ yd})\left(36\ \frac{\text{in}}{\text{yd}}\right) = 4320\text{ in}$$

$$D_{\text{photo}} = (9.144\text{ mm})\left(\frac{1\text{ in}}{25.4\text{ mm}}\right) = 0.36\text{ in}$$

The photo scale can then be determined.

$$\text{photo scale} = 1{:}\frac{D_{\text{ground}}}{D_{\text{photo}}} = 1{:}\frac{4320\text{ in}}{0.36\text{ in}} = 1{:}12{,}000$$

34 *The answer is (D).*

All measured distances must be reduced to sea level. Distances measured at an elevation above sea level are longer than their sea-level equivalents; a negative correction must therefore be applied. Likewise, all measured distances must be corrected by a grid scale factor before undertaking any computations on a state-plane coordinate system.

35 *The answer is (D).*

NGVD 29 is an older system and is less frequently used. NAD 27 and NAD 83 are not elevation datums but are state-plane coordinate systems. The most common elevation datum for modern surveys is NGVD 88.

36 *The answer is (B).*

The precision of the measurement is 0.5 in + 10 ppm of the length.

$$\text{precision} = 0.5\text{ in} + \left(\frac{10}{1{,}000{,}000}\right)L$$

$$= 0.5\text{ in} + \left(\frac{10}{1{,}000{,}000}\right)(1200\text{ ft})\left(12\ \frac{\text{in}}{\text{ft}}\right)$$

$$= 0.644\text{ in} \quad (0.6\text{ in})$$

37 *The answer is (B).*

Compute the x misclose.

$$x\text{ misclose} = \text{known } x - \text{computed } x = 2459.16\text{ ft} - 2459.38\text{ ft} = -0.22\text{ ft}$$

Compute the y misclose.

$$\begin{aligned} y \text{ misclose} &= \text{known } y - \text{computed } y \\ &= 5211.90 \text{ ft} - 5211.71 \text{ ft} \\ &= +0.19 \text{ ft} \end{aligned}$$

The x and y misclose are combined to form a diagonal linear misclose.

$$\begin{aligned} \text{linear misclose} &= \sqrt{(x \text{ mislose})^2 + (y \text{ misclose})^2} \\ &= \sqrt{(-0.22 \text{ ft})^2 + (0.19 \text{ ft})^2} \\ &= 0.2907 \text{ ft} \end{aligned}$$

The linear misclose is converted to an accuracy.

$$\begin{aligned} \text{accuracy} &= 1{:}\frac{\text{traverse length}}{\text{linear misclose}} \\ &= 1{:}\frac{3250 \text{ ft}}{0.2907 \text{ ft}} \\ &= 1{:}11{,}180 \quad (1{:}11{,}000) \end{aligned}$$

38 *The answer is (C).*

$$\begin{aligned} D_{\text{AB}} &= (145.76 \text{ ft}) \cos 2°20' \\ &= 145.639 \text{ ft} \\ D_{\text{BC}} &= (189.44 \text{ ft}) \cos 3°10' \\ &= 189.151 \text{ ft} \end{aligned}$$

Now calculate the distance AC using the law of cosines.

$$\begin{aligned} (\text{AC})^2 &= (\text{AB})^2 + (\text{BC})^2 - 2(\text{AB})(\text{BC}) \cos(180° - 30°) \\ &= (145.639 \text{ ft})^2 + (189.151 \text{ ft})^2 \\ &\quad - (2)(145.639 \text{ ft})(189.151 \text{ ft}) \cos 150° \\ &= 104,702.943 \text{ ft}^2 \end{aligned}$$

$$\begin{aligned} \text{AC} &= \sqrt{104{,}702.943 \text{ ft}^2} \\ &= 323.58 \text{ ft} \end{aligned}$$

39 *The answer is (C).*

$$\begin{aligned} L_c &= L + C_l \\ &= L + (T_l - T)kL \end{aligned}$$

The three corrected lengths are,

$$\begin{aligned} L_{C_1} &= 597.410 \text{ ft} + (83°\text{F} - 68°\text{F}) \\ &\quad \times \left(6.45 \times 10^{-6} \frac{1}{°\text{F}}\right) (597.410 \text{ ft}) \\ &= 597.468 \text{ ft} \\ L_{C_2} &= 597.320 \text{ ft} + (98°\text{F} - 68°\text{F}) \\ &\quad \times \left(6.45 \times 10^{-6} \frac{1}{°\text{F}}\right) (597.320 \text{ ft}) \\ &= 597.436 \text{ ft} \\ L_{C_3} &= 597.500 \text{ ft} + (49°\text{F} - 68°\text{F}) \\ &\quad \times \left(6.45 \times 10^{-6} \frac{1}{°\text{F}}\right) (597.410 \text{ ft}) \\ &= 597.427 \text{ ft} \end{aligned}$$

The mean of these three measurements will be the most probable length of the line.

$$\begin{aligned} \text{most probable length} &= \frac{L_{C_1} + L_{C_2} + L_{C_3}}{3} \\ &= \frac{\begin{array}{c} 597.468 \text{ ft} + 597.436 \text{ ft} \\ + 597.427 \text{ ft} \end{array}}{3} \\ &= 597.444 \text{ ft} \quad (597.44 \text{ ft}) \end{aligned}$$

40 *The answer is (B).*

$$\begin{aligned} L_{\text{sea level}} &= D\left(\frac{R}{R + \text{elev}}\right) \\ &= (2189.01 \text{ ft})\left(\frac{20{,}906{,}000 \text{ ft}}{20{,}906{,}000 \text{ ft} + 4100 \text{ ft}}\right) \\ &= 2188.58 \text{ ft} \end{aligned}$$

41 *The answer is (B).*

With the prismoidal formula, the middle section dimensions are obtained by simply averaging the two end section dimensions. A common mistake is to compute the two end cross-section areas and then average them.

42 *The answer is (A).*

$$\begin{aligned} V &= LWD \\ &= (80 \text{ ft})(60 \text{ ft})(5 \text{ ft})\left(\frac{1 \text{ yd}^3}{27 \text{ ft}^3}\right) \\ &= 888.89 \text{ yd}^3 \end{aligned}$$

Determine the total cost of excavation.

$$\begin{aligned} \text{CE}_{\text{total}} &= V(\text{CE}_{\text{yd}^3}) \\ &= (888.89\ \text{yd}^3)\left(\frac{\$2.00}{\text{yd}^3}\right) \\ &= \$1777.78 \end{aligned}$$

The dirt is moved 800 ft, or 80 sta. There is no charge for overhaul of the first five stations. Determine the total cost of overhaul.

$$\begin{aligned} \text{CO}_{\text{total}} &= V(\text{CO}_{\text{yd}^3}) \\ &= (888.89\ \text{yd}^3)\left(\frac{\$0.40}{\text{yd}^3/\text{sta}}\right)(8\ \text{sta} - 5\ \text{sta}) \\ &= \$1066.67 \end{aligned}$$

The total cost is then

$$\begin{aligned} C_{\text{total}} &= \text{CE}_{\text{total}} + \text{CO}_{\text{total}} \\ &= \$1777.78 + \$1066.67 \\ &= \$2844.45 \quad (\$2800) \end{aligned}$$

43 *The answer is (C).*

To orient a stereo model, the parameters to be solved are three rotations, three translations, and a uniform scale change. Two of the rotations (tip and tilt) can be solved using three height points, which must not be in a straight line. The azimuth rotation and scale change can be solved using two plan points. Once the model is rotated and scaled, any of the three height points and any of the two plan points can be used to solve for the three translations.

44 *The answer is (A).*

A change in the location of a legal boundary is known as a lot-line adjustment.

45 *The answer is (B).*

Boundary markers in California have taken many forms over the years. However, the current standard monument is the 1 in iron pipe with a license tag attached to the top.

46 *The answer is (C).*

First, determine the earth curvature correction and the refraction correction.

$$\begin{aligned} C_f &= 0.0239\left(\frac{D}{1000}\right)^2 \\ &= (0.0239)\left(\frac{3864.7\ \text{ft}}{1000}\right)^2 \\ &= 0.36\ \text{ft} \end{aligned}$$

$$\begin{aligned} R_f &= -0.0033\left(\frac{D}{1000}\right)^2 \\ &= (-0.0033)\left(\frac{3864.7\ \text{ft}}{1000}\right)^2 \\ &= -0.05\ \text{ft} \end{aligned}$$

The combined correction is then

$$\begin{aligned} C_c &= C_f + R_f \\ &= 0.36\ \text{ft} - 0.05\ \text{ft} \\ &= 0.31\ \text{ft} \end{aligned}$$

47 *The answer is (D).*

The height difference between the roof edge and the line of sight is 12.00 ft − 4.89 ft, or 7.11 ft. Therefore, the roof edge is 7.11 ft above the line of sight.

The height difference between the line of sight and the benchmark is equal to the rod reading of 7.90 ft. Therefore, the line of sight is 7.90 ft above the benchmark. The sum of these two height differences, 7.11 ft + 7.90 ft, or 15.01 ft, equals the height difference between the benchmark and the roof edge.

48 *The answer is (B).*

A 6 by 6 group of sections is called a township, and townships are located by a coordinate system where the axes are township and range. For example, "T3N, R5W" means that the township is the third township north of the baseline and five townships to the west of the principal meridian.

49 *The answer is (D).*

A property that is described in a legal description by "metes and bounds" is identified by the bearings and lengths of its boundaries.

50 *The answer is (D).*

The fifth measurement (178.74 ft) is clearly a blunder and should be omitted from the set of measurements. The standard error is computed from the remaining five measurements.

First, determine the mean of the five measurements.

$$\begin{aligned} x_m &= \frac{\Sigma x_n}{5} \\ &= \frac{\begin{array}{c}178.45 \text{ ft} + 178.48 \text{ ft} + 178.49 \text{ ft} \\ + 178.44 \text{ ft} + 178.47 \text{ ft}\end{array}}{5} \\ &= 178.466 \text{ ft} \end{aligned}$$

The standard deviation, σ, is then

$$\sigma = \sqrt{\frac{\Sigma(x_n - x_m)^2}{n_{\text{total}} - 1}}$$

$$\begin{aligned} x &= \sqrt{\frac{\begin{array}{l}(178.45 \text{ ft} - 178.466 \text{ ft})^2 \\ \quad + (178.48 \text{ ft} - 178.466 \text{ ft})^2 \\ \quad + (178.49 \text{ ft} - 178.466 \text{ ft})^2 \\ \quad + (178.44 \text{ ft} - 178.466 \text{ ft})^2 \\ \quad + (178.47 \text{ ft} - 178.466 \text{ ft})^2\end{array}}{5 - 1}} \\ &= \sqrt{\frac{0.0017}{4}} \\ &= 0.021 \text{ ft} \quad (0.02 \text{ ft}) \end{aligned}$$

Sample Exam 2
Problems

Answer Sheet

1. (A) (B) (C) (D)
2. (A) (B) (C) (D)
3. (A) (B) (C) (D)
4. (A) (B) (C) (D)
5. (A) (B) (C) (D)
6. (A) (B) (C) (D)
7. (A) (B) (C) (D)
8. (A) (B) (C) (D)
9. (A) (B) (C) (D)
10. (A) (B) (C) (D)
11. (A) (B) (C) (D)
12. (A) (B) (C) (D)
13. (A) (B) (C) (D)
14. (A) (B) (C) (D)
15. (A) (B) (C) (D)
16. (A) (B) (C) (D)
17. (A) (B) (C) (D)
18. (A) (B) (C) (D)
19. (A) (B) (C) (D)
20. (A) (B) (C) (D)
21. (A) (B) (C) (D)
22. (A) (B) (C) (D)
23. (A) (B) (C) (D)
24. (A) (B) (C) (D)
25. (A) (B) (C) (D)
26. (A) (B) (C) (D)
27. (A) (B) (C) (D)
28. (A) (B) (C) (D)
29. (A) (B) (C) (D)
30. (A) (B) (C) (D)
31. (A) (B) (C) (D)
32. (A) (B) (C) (D)
33. (A) (B) (C) (D)
34. (A) (B) (C) (D)
35. (A) (B) (C) (D)
36. (A) (B) (C) (D)
37. (A) (B) (C) (D)
38. (A) (B) (C) (D)
39. (A) (B) (C) (D)
40. (A) (B) (C) (D)
41. (A) (B) (C) (D)
42. (A) (B) (C) (D)
43. (A) (B) (C) (D)
44. (A) (B) (C) (D)
45. (A) (B) (C) (D)
46. (A) (B) (C) (D)
47. (A) (B) (C) (D)
48. (A) (B) (C) (D)
49. (A) (B) (C) (D)
50. (A) (B) (C) (D)

1 *(4 points)*

Control surveys that are used for aerial mapping projects are usually based on which survey method?

(A) total station
(B) global positioning system
(C) triangulation
(D) stadia

2 *(4 points)*

The global positioning system (GPS) survey method used most frequently in construction surveying is

(A) real-time kinematic
(B) kinematic
(C) static
(D) fast static

3 *(6 points)*

The following parameters define a horizontal curve.

deflection angle	12°00′00″
curve radius	2200.00 ft
PI (point of intersection) stationing	sta 20+15.9

The length of the curve is

(A) 230.38 ft
(B) 418.88 ft
(C) 460.54 ft
(D) 460.77 ft

4 *(4 points)*

The distinctive feature of a tilting level is that it

(A) does not need to be leveled for each sight
(B) has a sensitive bubble that is used to level the line of sight for each observation
(C) has to be leveled only once, using a circular bubble
(D) is attached to an instrument that is leveled for each observation using footscrews

5 *(4 points)*

A registered civil engineer may establish property boundaries if the engineer

(A) was registered prior to January 1, 1992
(B) has a minimum of eight years of land surveying experience under the supervision of licensed surveyors and has passed the LSIT exam
(C) was registered prior to January 1, 1972
(D) has a bachelor's degree in land surveying

6 *(4 points)*

Maps that are compiled from an aerial survey may be authorized by

(A) a licensed land surveyor
(B) a civil engineer registered before January 1, 1982
(C) any registered civil engineer
(D) all of the above

7 *(8 points)*

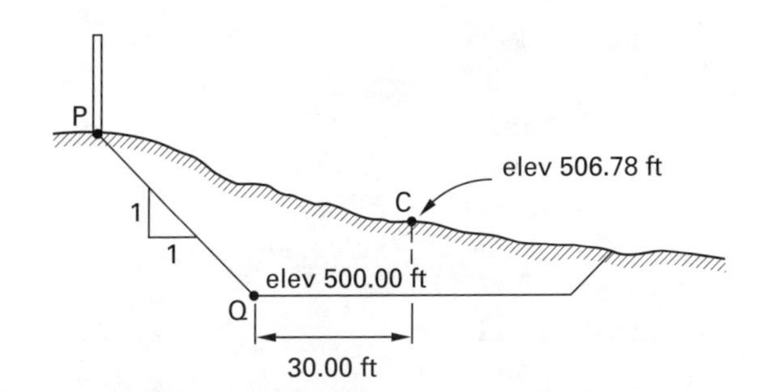

For the cross section shown, the terrain has a 2% grade. The elevation of the slope stake at P is

(A) 506.78 ft
(B) 506.92 ft
(C) 507.24 ft
(D) 507.53 ft

8 *(6 points)*

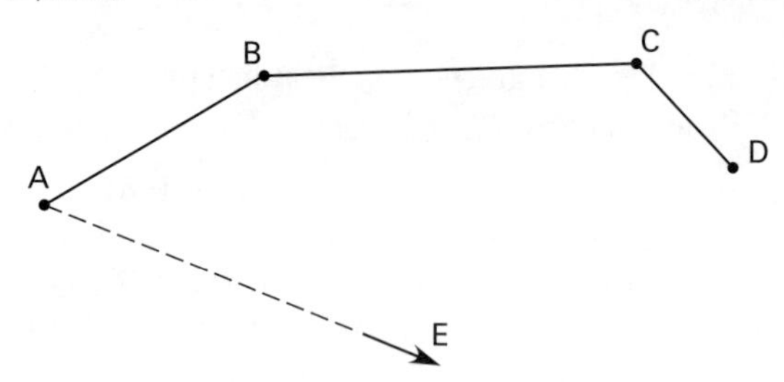

In the open traverse shown,

azimuth AE	111°47′36″
interior angle EAB	51°10′19″
deflection angle at B	28°00′01″
interior angle BCD	132°00′00″

The azimuth of line CD is

(A) 80°37′16″
(B) 136°37′18″
(C) 238°57′56″
(D) 260°37′16″

9 *(6 points)*

The observed angles of a four-course closed traverse are

angle A	88°47′30″
angle B	110°15′50″
angle C	79°00′10″
angle D	81°55′50″

The balanced angle D is

(A) 81°55′10″
(B) 81°55′40″
(C) 81°56′00″
(D) 81°56′40″

10 *(4 points)*

The term RPSS on a construction stake stands for

(A) reference point side stake
(B) reference point slope stake
(C) right point side slope
(D) rough point side stake

11 *(6 points)*

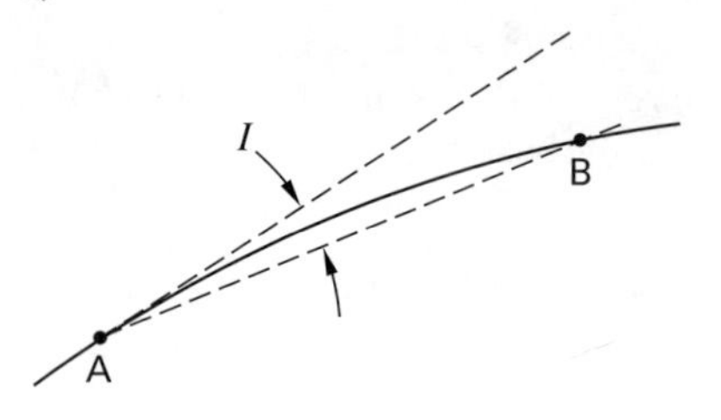

AB is a circular curve on a horizontal alignment. The following parameters apply.

stationing at A	sta 15+45.8
stationing at B	sta 19+95.8
radius	2500 ft

The deflection angle, I, is

(A) 5°02′31″
(B) 5°09′24″
(C) 5°09′33″
(D) 5°15′46″

12 *(6 points)*

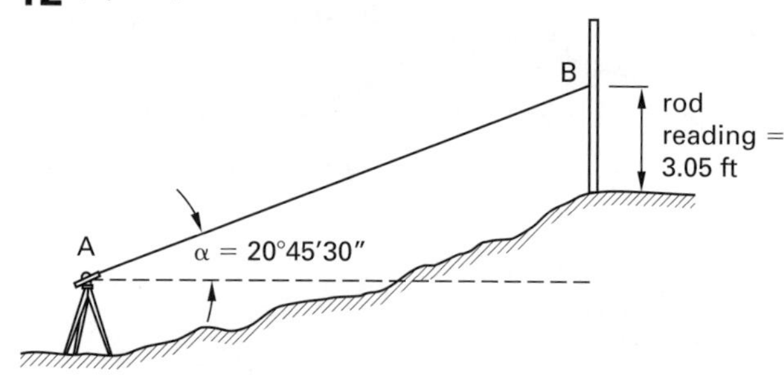

For the slope shown, the following parameters apply.

height of instrument at A	4.52 ft
elevation of A	497.26 ft
slope distance (SD) (measured along line of sight)	187.44 ft

From the given measurements, the elevation of stake B is

(A) 565.16 ft
(B) 569.78 ft
(C) 571.26 ft
(D) 575.89 ft

13 *(6 points)*

A 300 ft steel tape is used to place two stakes: one at the 0 ft mark and one at the 300 ft mark. The air temperature is 93°F. If the standard temperature of the steel tape used is 68°F, the true distance between the stakes is

(A) 299.95 ft
(B) 299.98 ft
(C) 300.02 ft
(D) 300.05 ft

14 *(8 points)*

point	BS (ft)	FS (ft)	prelim. elevation (ft)	final elevation (ft)
BM1	3.57	–	–	100.00
	2.56	6.61	–	–
TP1	4.91	5.89	–	–
	3.33	4.67	–	–
BM2	–	6.72	–	90.60

Using data given in the table, the adjusted (balanced) elevation of TP1 is

(A) 93.57 ft
(B) 93.63 ft
(C) 93.69 ft
(D) 106.37 ft

15 *(8 points)*

At a setup during profile leveling, three crosshair readings were recorded at points A and B, as shown.

point	A (ft)	B (ft)
top	5.15	6.30
mid	5.00	5.92
bottom	4.85	5.54

The grade of line AB is

(A) −2%
(B) −1%
(C) 1%
(D) 2%

16 *(6 points)*

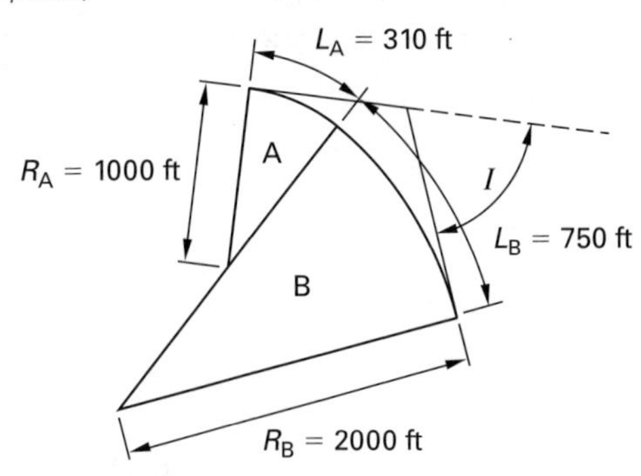

The total intersection angle, I, for the compound curve shown is

(A) 39°09′57″
(B) 39°10′26″
(C) 39°14′51″
(D) 39°25′16″

17 *(8 points)*

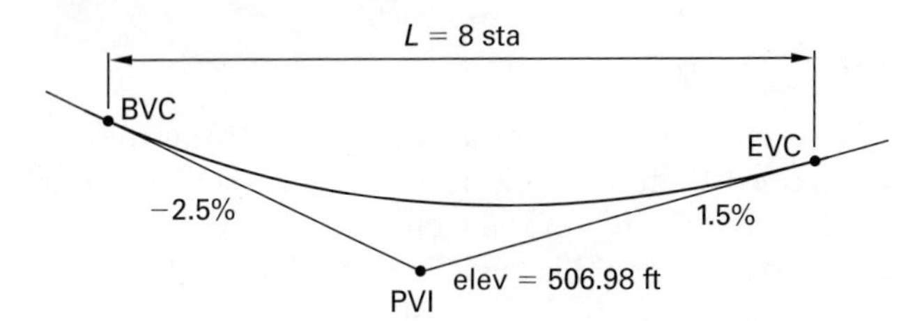

The equation of the curve shown is

(A) $y = 516.98 \text{ ft} - 2.5x + 0.5x^2$
(B) $y = 496.98 \text{ ft} - 1.5x + 0.25x^2$
(C) $y = 496.98 \text{ ft} + 2.5x + 0.5x^2$
(D) $y = 516.98 \text{ ft} - 2.5x + 0.25x^2$

18 *(6 points)*

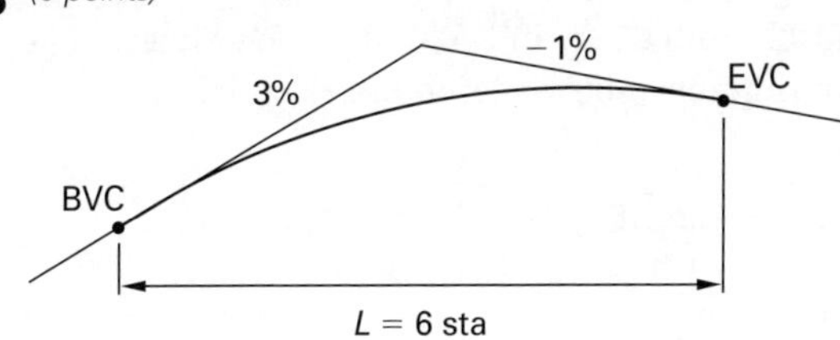

The tangent offset to the vertical curve at $x = 2.00$ sta from the beginning of the vertical curve (BVC) is

(A) −2.00 ft
(B) −1.32 ft
(C) 1.32 ft
(D) 2.00 ft

19 *(6 points)*

A backsight of 1.67 ft is read to a benchmark with a known elevation of 378.54 ft. The foresight is to the rod resting on an exposed horizontal pipe, and the foresight reading is 8.06 ft. The pipe has an inside diameter of 5 ft 4 in, and the thickness of the pipe casing is 1 in. The elevation of the center of the pipe is

(A) 369.40 ft
(B) 377.57 ft
(C) 379.51 ft
(D) 390.35 ft

20 *(4 points)*

The distance from a control station to a distant mountain scaled from a map is 18.4 mi. A vertical angle is measured to the top of the mountain. The earth curvature correction for this angle line is most nearly

(A) 20 ft
(B) 23 ft
(C) 200 ft
(D) 230 ft

21 *(4 points)*

During a profile leveling run, the backsight to point A (elevation 500.00 ft) is 4.77 ft. The foresights to points B and C are 3.65 ft and 5.89 ft, respectively. The elevation difference between points B and C is

(A) −3.24 ft
(B) −2.24 ft
(C) 2.24 ft
(D) 3.24 ft

22 *(6 points)*

A traverse closes to 1/84,000. The lengths of the traverse sides are given in the table.

traverse side	length (ft)
AB	10,256
BC	8234
CD	4744
DA	12,399

The traverse misclose in terms of feet on the ground is

(A) 0.18 ft
(B) 0.38 ft
(C) 0.42 ft
(D) 0.65 ft

23 *(4 points)*

A boundary survey displayed on a map that is officially approved by a county is known as

(A) an official map
(B) a legal map
(C) a recorded map
(D) a county map

24 *(4 points)*

The (x, y) coordinates, in feet, of two points A and B are (101.56 ft, 556.23 ft) and (637.89 ft, 15.33 ft), respectively, as shown. The radius of the circular arc (center at B) is 700.00 ft.

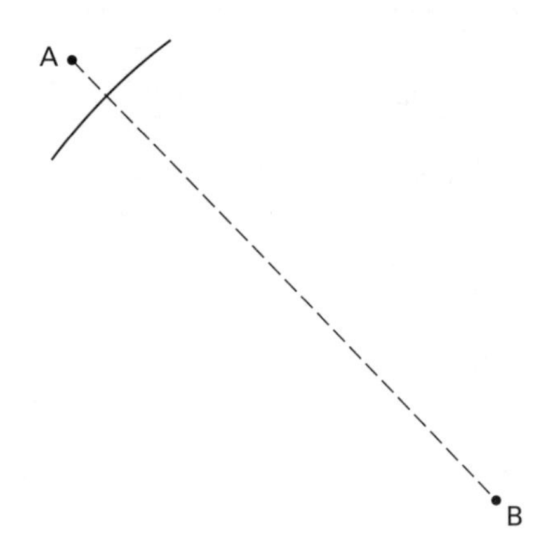

The shortest distance from A to the arc is

(A) 7.86 ft
(B) 61.7 ft
(C) 87.9 ft
(D) 761.7 ft

25 *(6 points)*

Measured highway cross sections S_1 and S_2 are 55.0 ft apart and have areas of 127.22 ft^2 and 187.56 ft^2, respectively. The volume between the sections generated by the end-area method is most nearly

(A) 191 yd^3
(B) 321 yd^3
(C) 642 yd^3
(D) 8660 yd^3

26 *(6 points)*

The end cross-section areas on a 50.0 ft length of highway are fill 98.24 ft^2 and fill 110.43 ft^2. A prismoidal volume is required, and the computed midway section cross-section area is fill 106.59 ft^2. Using the prismoidal formula, the volume of fill between the end sections is

(A) 190 yd^3
(B) 195 yd^3
(C) 196 yd^3
(D) 216 yd^3

27 *(4 points)*

In a horizontal circular curve with an intersection angle of 10°30′00″ and a radius of 2200.00 ft, the tangent distance is

(A) 183.77 ft
(B) 202.15 ft
(C) 202.30 ft
(D) 407.75 ft

28 *(6 points)*

On a vertical curve with grades of −2% and +3% and a length of 8.0 sta, the station of the lowest point on the curve is

(A) sta 2.4
(B) sta 3.2
(C) sta 4.8
(D) sta 5.4

29 *(6 points)*

A line of levels is run from a benchmark START to a point END. The elevation of END as computed from the leveling is 3679.65 ft. The known elevation of END is 3679.39 ft. The computed elevation of a point MID (midway between START and END) is 3500.00 ft. The adjusted (balanced) elevation of MID is

(A) 3499.74 ft
(B) 3499.87 ft
(C) 3500.13 ft
(D) 3500.26 ft

30 *(6 points)*

The distance and bearing from a point A to a point B are 130.65 ft and N 18°56′00″ W. If the (x, y) coordinates of point A are (310.00 ft, 275.00 ft), the (x, y) coordinates of point B are

(A) (258.54 ft, 261.99 ft)
(B) (267.61 ft, 398.58 ft)
(C) (352.39 ft, 398.58 ft)
(D) (433.58 ft, 232.61 ft)

31 *(4 points)*

In a public lands survey, the azimuth, α, from the NW corner of section 10 to the SE corner of section 11 is most nearly

(A) 26°33′54″
(B) 116°33′54″
(C) 180°00′00″
(D) 243°26′06″

32 *(6 points)*

zenith angle PQ = 93°44′20″
slope distance PQ = 265.95 ft

From the given measurements, the horizontal distance PQ is

(A) 15.97 ft
(B) 17.34 ft
(C) 265.38 ft
(D) 265.47 ft

33 *(4 points)*

point	BS (ft)	HI (ft)	FS (ft)	IFS (ft)	elevation (ft)
A	4.87	–	–	–	100.00
B	–	–	–	6.87	–
C	–	–	–	5.91	–
D	–	–	–	4.77	–

Given the observations of profile leveling in the table, the elevation of point D is

(A) 87.32 ft
(B) 99.90 ft
(C) 100.10 ft
(D) 112.68 ft

34 *(6 points)*

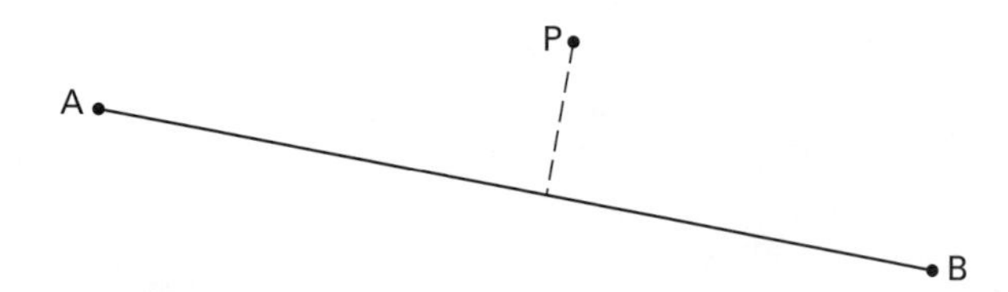

The bearing of line BA is N 79°34′50″ W. The azimuth of the shortest line from point P to line AB is

(A) 10°25′10″
(B) 169°34′50″
(C) 190°25′10″
(D) 349°34′50″

35 *(4 points)*

The area of a section in the U.S. Public Lands System is

(A) 100 ac
(B) 440 ac
(C) 640 ac
(D) 1000 ac

36 *(4 points)*

The projection of the California State Plane System is

(A) transverse mercator
(B) Albers equal area
(C) orthographic
(D) Lambert conformal conic

37 *(4 points)*

The standard method of computing a volume of a proposed dam basin from a contour map is to

(A) run longitudinal sections across the basin, determine the section areas, and use the end-area method to compute the volume
(B) mark random points on each contour and use these to form a triangulated irregular network (TIN) file from which the volume may be determined
(C) superimpose a grid over the basin, compute the volume below each grid square, and sum these volumes to obtain a total volume of the basin
(D) determine the area within each contour covering the basin and use the end-area method for the volume computation

38 *(4 points)*

If the forward overlap on a pair of adjacent aerial photos is 60% and the photo scale is 1 in:400 ft, the distance on the ground between the photo centers is

(A) 160 ft
(B) 240 ft
(C) 1440 ft
(D) 2160 ft

39 *(4 points)*

A final map is compiled when

(A) final earthwork quantities have been determined
(B) a parcel of land is subdivided into five or more smaller parcels
(C) a photogrammetric map is fully field checked
(D) a boundary dispute is resolved by field survey

40 *(4 points)*

(latitude, departure) from point A to B
$= (78.34 \text{ ft}, 64.90 \text{ ft})$

(latitude, departure) from point B to C
$= (-176.38 \text{ ft}, 94.67 \text{ ft})$

From the given latitude and departure data, distance AC is most nearly

(A) 100 ft
(B) 190 ft
(C) 260 ft
(D) 3000 ft

41 *(4 points)*

A POB is the starting point on a

(A) legal description
(B) traverse
(C) highway realignment
(D) public lands township

42 *(4 points)*

A digital alternative to a contour map is called a

(A) digital terrain model
(B) digital elevation model
(C) TIN file
(D) triangulated network

43 *(4 points)*

The datum for height above sea level prior to 1991 was

(A) NAD 27
(B) NAD 83
(C) NGVD 29
(D) NGVD 88

44 *(4 points)*

The most common side overlap used in aerial mapping is

(A) 25%
(B) 30%
(C) 60%
(D) 70%

45 *(4 points)*

A parcel map is used when

(A) a boundary dispute has to be referred to a court
(B) an old property is missing all its corner points and is established entirely from adjacent parcels
(C) two adjacent properties are merged into a single new parcel
(D) a property is subdivided into four or fewer parcels

46 *(4 points)*

The midpoint of a section boundary in the Public Lands System is called a

(A) quarter corner
(B) half corner
(C) midpoint
(D) township corner

47 *(4 points)*

When measuring the elevations of the terrain along a cross section on a new highway alignment, the height of instrument is

(A) measured with a tape
(B) determined from the backsight to a point with known elevation
(C) determined from a foresight
(D) not required

48 *(4 points)*

In a leveling field book, the correct procedure for checking the computation of elevations is to

(A) compute the elevation differences between successive points and compare these with the backsights minus the foresights
(B) compare the average of all the computed elevations with the sum of the backsights minus the sum of the foresights
(C) repeat the computations
(D) sum the backsights and subtract the sum of the foresights, and compare this value with the elevation difference between the end point and the start point

49 *(4 points)*

The original basic unit of measure in the Public Lands System is the

(A) U.S. survey foot
(B) chain
(C) link
(D) meter

50 *(4 points)*

The U.S. survey foot is linked to which exact conversion factor?

(A) 3937/1200
(B) 3.2808
(C) 3.048
(D) 25.4

Sample Exam 2
Solutions

Answer Key

No.	Answer	Point Value
1.	B	4
2.	A	4
3.	D	6
4.	B	4
5.	C	4
6.	D	4
7.	D	8
8.	B	6
9.	C	6
10.	B	4
11.	B	6
12.	A	6
13.	D	6
14.	C	8
15.	D	8
16.	C	6
17.	D	8
18.	B	6
19.	A	6
20.	D	4
21.	C	4
22.	C	6
23.	C	4
24.	B	4
25.	B	6
26.	C	6
27.	B	4
28.	B	6
29.	B	6
30.	B	6
31.	B	4
32.	C	6
33.	C	4
34.	C	6
35.	C	4
36.	D	4
37.	D	4
38.	C	4
39.	B	4
40.	B	4
41.	A	4
42.	B	4
43.	C	4
44.	B	4
45.	D	4
46.	A	4
47.	B	4
48.	D	4
49.	B	4
50.	A	4

Perfect Score = 250 points
Point Total: ____________

Note: A point total of greater than 150 can be considered a passing score on this sample exam.

1 *The answer is (B).*

Aerial mapping control is almost always surveyed using the global positioning system (GPS). Aerial mapping projects usually cover large areas, and GPS is particularly suited to this application, since intervisibility between points is not required. Triangulation and stadia are now obsolete methods, and total station surveying is best suited to smaller projects or to projects in built-up areas where GPS might not be practical because of interference by tall buildings and trees.

2 *The answer is (A).*

The static and fast-static GPS point measurement methods require a minimum of 5 min per point, and are therefore too slow for construction staking. The kinematic survey method is much faster in its point determination, but it has limitations relating to keeping a lock on the satellites. The best method for construction surveys is real-time kinematic (RTK), where the rover receiver is connected to the base station via a radio link and point determination is instantaneous.

3 *The answer is (D).*

Determine the length of the curve.

$$\begin{aligned} L &= RI\left(\frac{\pi}{180°}\right) \\ &= (2200.00 \text{ ft})(12°)\left(\frac{\pi}{180°}\right) \\ &= 460.77 \text{ ft} \end{aligned}$$

4 *The answer is (B).*

A tilting level has a separate bubble that is attached to the instrument. This bubble is often in the form of a split bubble that must be leveled for each line of sight.

5 *The answer is (C).*

A registered civil engineer who was registered prior to January 1, 1982, is exempt from land surveying license requirements. The same applies if the registration was prior to January 1, 1972. The LSIT (Land Surveyor-in-Training) is only a preliminary examination. A registered civil engineer with a bachelor's degree in land surveying would still need to pass the Professional Land Surveyor (PLS) Examination in order to be licensed as a land surveyor in California.

6 *The answer is (D).*

Aerial survey maps can be approved by any licensed civil engineer or licensed land surveyor.

7 *The answer is (D).*

The elevation of a point is determined from

$$y = y_{\text{known}} + gx$$

Assuming the horizontal distance component QP to be x (in ft),

$$\begin{aligned} y_{\text{P}} &\text{ computed from } y_{\text{Q}} \\ &= 500.00 \text{ ft} + x \\ y_{\text{P}} &\text{ computed from } y_{\text{C}} \\ &= 506.78 \text{ ft} + (0.02)(x + 30 \text{ ft}) \end{aligned}$$

Equating the two point elevation equations,

$$\begin{aligned} 500.00 \text{ ft} + x &= 506.78 \text{ ft} + 0.02x + 0.60 \text{ ft} \\ 0.98x &= 7.38 \text{ ft} \\ x &= 7.53 \text{ ft} \end{aligned}$$

Since the slope of PQ is 1:1,

$$\begin{aligned} y_{\text{P}} &= 500.00 \text{ ft} + x = 500.00 \text{ ft} + 7.53 \text{ ft} \\ &= 507.53 \text{ ft} \end{aligned}$$

This can be verified by computing the grade of terrain CP using the computed elevation of point P.

8 *The answer is (B).*

$$\begin{aligned} \alpha_{\text{AB}} &= \alpha_{\text{AE}} - \text{angle EAB} \\ &= 111°47'36'' - 51°10'19'' \\ &= 60°37'17'' \\ \alpha_{\text{BC}} &= \alpha_{\text{AB}} + \text{deflection angle B} \\ &= 60°37'17'' + 28°00'01'' \\ &= 88°37'18'' \\ \alpha_{\text{CD}} &= \alpha_{\text{BC}} + 180°00'00'' - \text{angle BCD} \\ &= 88°37'18'' + 180° - 132°00'00'' \\ &= 136°37'18'' \end{aligned}$$

9 *The answer is (C).*

The sum of the four angles is 359°59′20″. The corrected sum should be 360°00′00″.

$$\begin{aligned}\text{misclose} &= 360^\circ - \sum 4 \text{ angles}\\ &= 360^\circ - 359^\circ59'20''\\ &= 40''\\ \text{correction per angle} &= \frac{\text{misclose}}{\text{no. of angles}}\\ &= \frac{40''}{4}\\ &= 10''\\ \text{balanced angle D} &= 81^\circ55'50'' + \text{correction}\\ &= 81^\circ55'50'' + 10''\\ &= 81^\circ56'00''\end{aligned}$$

10 *The answer is (B).*

The letters RPSS, which stand for reference point slope stake, would be written on a witness stake to indicate that the data on the stake refers to a slope stake some distance away (e.g., 10 ft). The stake next to the witness stake would be a reference point.

11 *The answer is (B).*

The deflection angle is

$$\begin{aligned}I &= \tfrac{1}{2}(\text{central angle subtended by AB}) = \frac{L}{2R}\\ &= \frac{\text{sta at B} - \text{sta at A}}{2R}\\ &= \left(\frac{19.958 \text{ sta} - 15.458 \text{ sta}}{(2)(2500 \text{ ft})}\right)\left(\frac{180^\circ}{\pi \text{ rad}}\right)\\ &= 5.1566^\circ \quad (5^\circ09'24'')\end{aligned}$$

12 *The answer is (A).*

$$\begin{aligned}y_\text{B} &= y_\text{A} + \text{HI} + \text{SD}\sin\alpha - \text{rod reading}\\ &= 497.26 \text{ ft} + 4.52 \text{ ft} + (187.44 \text{ ft})\sin 20^\circ45'30''\\ &\quad - 3.05 \text{ ft}\\ &= 565.16 \text{ ft}\end{aligned}$$

13 *The answer is (D).*

The difference between the standard temperature of the steel tape and the air temperature is

$$\Delta T = 93^\circ\text{F} - 68^\circ\text{F} = 25^\circ\text{F}$$

The coefficient of thermal expansion of steel, α, is 6.45 $\times 10^{-6}/^\circ$F. Therefore,

$$\begin{aligned}\text{correction } \Delta L &= \alpha L \Delta T\\ &= \left(6.45 \times 10^{-6} \frac{1}{^\circ\text{F}}\right)(300.00 \text{ ft})(25^\circ\text{F})\\ &= 0.05 \text{ ft}\end{aligned}$$

At the colder standard temperature, the tape will contract by 0.05 ft. Therefore, the 300.00 ft mark will be on the side of the stake toward the 0 ft mark. The measured distance between the stakes at 68°F will thus be 300.05 ft.

14 *The answer is (C).*

The unadjusted elevation is determined from

$$y_\text{unadj} = y_\text{BM1} + \text{BS1} - \text{FS1} + \text{BS2} - \text{FS2} + \ldots$$

Therefore, the unadjusted elevations for test point 1 and benchmark 2 are

$$\begin{aligned}y_\text{TP1,unadj} &= 100.00 \text{ ft} + 3.57 \text{ ft} - 6.61 \text{ ft}\\ &\quad + 2.56 \text{ ft} - 5.89 \text{ ft}\\ &= 93.63 \text{ ft}\\ y_\text{BM2,unadj} &= 100.00 \text{ ft} + 3.57 \text{ ft} - 6.61 \text{ ft}\\ &\quad + 2.56 \text{ ft} - 5.89 \text{ ft} + 4.91 \text{ ft}\\ &\quad - 4.67 \text{ ft} + 3.33 \text{ ft} - 6.72 \text{ ft}\\ &= 90.48 \text{ ft}\end{aligned}$$

The misclose is

$$\begin{aligned}\text{misclose} &= y_\text{BM2,final} - y_\text{BM2,unadj}\\ &= 90.60 \text{ ft} - 90.48 \text{ ft}\\ &= 0.12 \text{ ft}\end{aligned}$$

Since TP1 is the midpoint of the leveling,

$$\begin{aligned}y_\text{TP1,adj} &= y_\text{TP1,unadj} + \left(\tfrac{1}{2}\right)\text{misclose}\\ &= 93.63 \text{ ft} + \left(\tfrac{1}{2}\right)(0.12 \text{ ft})\\ &= 93.69 \text{ ft}\end{aligned}$$

15 *The answer is (D).*

The distance from the instrument to point A is

$$\begin{aligned} x_{\text{A}} &= 100\begin{pmatrix}\text{top reading} \\ \quad -\text{bottom reading}\end{pmatrix} \\ &= (100)(5.15\ \text{ft} - 4.85\ \text{ft}) \\ &= 30.0\ \text{ft} \end{aligned}$$

The distance from the instrument to point B is

$$\begin{aligned} x_{\text{B}} &= 100\begin{pmatrix}\text{top reading} \\ \quad -\text{bottom reading}\end{pmatrix} \\ &= (100)(6.30\ \text{ft} - 5.54\ \text{ft}) \\ &= 76.0\ \text{ft} \end{aligned}$$

It follows that distance AB is

$$\begin{aligned} x_{\text{AB}} &= x_{\text{B}} - x_{\text{A}} = 76.0\ \text{ft} - 30.0\ \text{ft} \\ &= 46.0\ \text{ft} \\ \Delta y_{\text{AB}} &= (\text{mid reading to B}) - (\text{mid reading to A}) \\ &= 5.92\ \text{ft} - 5.00\ \text{ft} \\ &= 0.92\ \text{ft} \end{aligned}$$

The grade of line AB is

$$\begin{aligned} g_{\text{AB}} &= \frac{\Delta y_{\text{AB}}}{x_{\text{AB}}} \times 100\% \\ &= \frac{0.92\ \text{ft}}{46.0\ \text{ft}} \times 100\% \\ &= 2\% \end{aligned}$$

16 *The answer is (C).*

$$\begin{aligned} \text{central angle} &= \frac{L}{R} \\ \text{central angle of A} &= \frac{310\ \text{ft}}{1000\ \text{ft}} \\ &= 0.310\ \text{rad} \\ \text{central angle of B} &= \frac{750\ \text{ft}}{2000\ \text{ft}} \\ &= 0.375\ \text{rad} \end{aligned}$$

The total intersection angle is

$$\begin{aligned} I &= \text{central angle of A} + \text{central angle of B} \\ &= (0.310\ \text{rad} + 0.375\ \text{rad})\left(\frac{180^\circ}{\pi\ \text{rad}}\right) \\ &= 39.2476^\circ \quad (39^\circ 14' 51'') \end{aligned}$$

17 *The answer is (D).*

The basic equation of a parabolic vertical curve is

$$\begin{aligned} y &= y_{\text{BVC}} + g_1 x + \left(\frac{r}{2}\right)x^2 \\ y_{\text{BVC}} &= y_{\text{PVI}} - g_1\left(\frac{L}{2}\right) \\ &= 506.98\ \text{ft} - (-2.5\%)\left(\frac{8\ \text{sta}}{2}\right) \\ &= 516.98\ \text{ft} \\ r &= \frac{g_2 - g_1}{L} = \frac{1.5\% - (-2.5\%)}{8\ \text{sta}} \\ &= 0.5\ \%/\text{sta} \end{aligned}$$

Substituting into the curve equation,

$$\begin{aligned} y &= 516.98\ \text{ft} + (-2.5x) + \left(\frac{0.5\ \frac{\%}{\text{sta}}}{2}\right)x^2 \\ &= 516.98\ \text{ft} - 2.5x + 0.25x^2 \end{aligned}$$

18 *The answer is (B).*

The rate of change of grade is

$$r = \frac{g_2 - g_1}{L} = \frac{-1\% - 3\%}{6.00\ \text{sta}} = -0.66\ \%/\text{sta}$$

This tangent offset is

$$\begin{aligned} y' &= \left(\frac{r}{2}\right)x^2 \\ &= \left(\frac{-0.66\frac{\%}{\text{sta}}}{2}\right)(2\ \text{sta})^2 \\ &= -1.32\ \text{ft} \end{aligned}$$

19 *The answer is (A).*

The elevation of the bottom of the rod (top surface of the pipe) is

$$\begin{aligned} y_{\text{rod bottom}} &= \text{BS} - \text{FS} + y_{\text{BM}} \\ &= 1.67\ \text{ft} - 8.06\ \text{ft} + 378.54\ \text{ft} \\ &= 372.15\ \text{ft} \end{aligned}$$

The height difference between the center and the outside of the pipe is

$$\Delta y = \frac{5 \text{ ft } 4 \text{ in}}{2} + 1 \text{ in} = 2 \text{ ft } 9 \text{ in}$$

The center of the pipe is therefore 2 ft 9 in (or 2.75 ft) below the top surface of the pipe.

$$\begin{aligned} y_{\text{pipe center}} &= y_{\text{rod bottom}} - \Delta y \\ &= 372.15 \text{ ft} - 2.75 \text{ ft} \\ &= 369.40 \text{ ft} \end{aligned}$$

20 *The answer is (D).*

The distance in feet is

$$x = (18.4 \text{ mi})\left(5280 \ \frac{\text{ft}}{\text{mi}}\right) = 97{,}152 \text{ ft}$$

The earth curvature correction can then be determined.

$$\begin{aligned} C_f &= (0.0239)\left(\frac{x}{1000}\right)^2 \\ &= (0.0239)\left(\frac{97{,}152 \text{ ft}}{1000}\right)^2 \\ &= 226 \text{ ft} \quad (230 \text{ ft}) \end{aligned}$$

21 *The answer is (C).*

$$\begin{aligned} y &= y_{\text{known}} + \text{BS} - \text{FS} \\ y_{\text{B}} &= 500.00 \text{ ft} + 4.77 \text{ ft} - 3.65 \text{ ft} \\ &= 501.12 \text{ ft} \\ y_{\text{C}} &= 500.00 \text{ ft} + 4.77 \text{ ft} - 5.89 \text{ ft} \\ &= 498.88 \text{ ft} \\ \Delta y_{\text{BC}} &= 501.12 \text{ ft} - 498.88 \text{ ft} \\ &= 2.24 \text{ ft} \end{aligned}$$

22 *The answer is (C).*

$$\frac{1}{84{,}000} = \frac{\text{misclose}}{\sum \text{side lengths}}$$

Rearranging the equation gives,

$$\frac{\sum \text{side lengths}}{84{,}000} = \text{misclose}$$

$$\begin{aligned} \sum \text{side lengths} &= 10{,}256 \text{ ft} + 8234 \text{ ft} \\ &\quad + 4744 \text{ ft} + 12{,}399 \text{ ft} \\ &= 35{,}633 \text{ ft} \\ \text{misclose} &= \frac{35{,}633 \text{ ft}}{84{,}000} \\ &= 0.42 \text{ ft} \end{aligned}$$

23 *The answer is (C).*

Boundary maps that are approved by a county are known as recorded maps.

24 *The answer is (B).*

Distance AB is

$$\begin{aligned} x_{\text{AB}} &= \sqrt{(x_{\text{A}} - x_{\text{B}})^2 + (y_{\text{A}} - y_{\text{B}})^2} \\ &= \sqrt{\begin{array}{l}(101.56 \text{ ft} - 637.89 \text{ ft})^2 \\ \quad + (556.23 \text{ ft} - 15.33 \text{ ft})^2\end{array}} \\ &= 761.72 \text{ ft} \end{aligned}$$

Therefore,

$$\begin{aligned} \text{shortest distance A to arc} &= x_{\text{AB}} - R \\ &= 761.72 \text{ ft} - 700.00 \text{ ft} \\ &= 61.72 \text{ ft} \quad (61.7 \text{ ft}) \end{aligned}$$

25 *The answer is (B).*

The end-area volume is

$$\begin{aligned} V &= \left(\frac{A_{S_1} + A_{S_2}}{2}\right) L \\ &= \left(\frac{127.22 \text{ ft}^2 + 187.56 \text{ ft}^2}{2}\right)\left(\frac{55.0 \text{ ft}}{27 \ \frac{\text{ft}^3}{\text{yd}^3}}\right) \\ &= 320.61 \text{ yd}^3 \quad (321 \text{ yd}^3) \end{aligned}$$

26 *The answer is (C).*

The volume of fill is

$$V = \frac{L(A_1 + 4A_m + A_2)}{6}$$
$$= \frac{(50.0\text{ ft})(98.24\text{ ft}^2 + (4)(106.59\text{ ft}^2) + 110.43\text{ ft}^2)}{(6)\left(27\ \frac{\text{ft}^3}{\text{yd}^3}\right)}$$
$$= 196\text{ yd}^3$$

27 *The answer is (B).*

The tangent distance is

$$T = R\tan\frac{I}{2}$$
$$= (2200.00\text{ ft})\tan\frac{10°30'}{2}$$
$$= 202.15\text{ ft}$$

28 *The answer is (B).*

The station of the lowest point on the curve, x, is

$$x = -\frac{g_1}{r}$$

Use the following formula to determine the rate of change of grade.

$$r = \frac{g_2 - g_1}{L} = \frac{3\ \frac{\%}{\text{sta}} - \left(-2\ \frac{\%}{\text{sta}}\right)}{8.0\text{ sta}}$$
$$= 0.625\ \%/\text{sta}^2$$

Therefore,

$$x = -\frac{-2\ \frac{\%}{\text{sta}}}{0.625\ \frac{\%}{\text{sta}^2}} = \text{sta } 3.2$$

29 *The answer is (B).*

The misclose at END is

$$\text{misclose} = y_{\text{END,known}} - y_{\text{END,computed}}$$
$$= 3679.39\text{ ft} - 3679.65\text{ ft}$$
$$= -0.26\text{ ft}$$

Since MID is halfway along the leveling route, the correction to the computed elevation of MID will be equal to half the misclose.

$$y_{\text{MID,adjusted}} = 3500\text{ ft} + \tfrac{1}{2}\text{misclose}$$
$$= 3500\text{ ft} + \left(\tfrac{1}{2}\right)(-0.26\text{ ft})$$
$$= 3499.87\text{ ft}$$

30 *The answer is (B).*

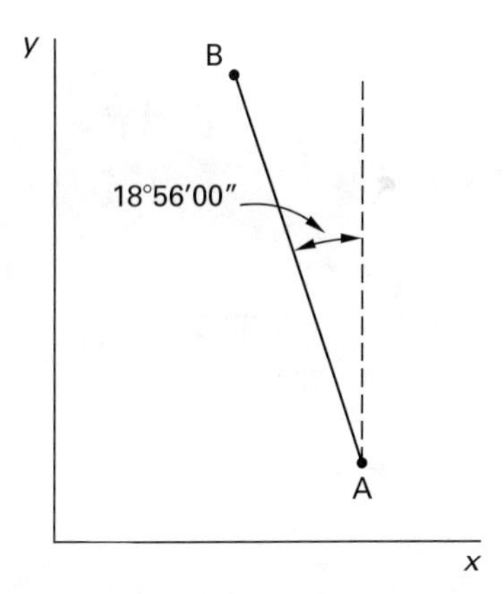

The azimuth of line AB, α, is

$$\alpha = 360°00'00'' - 18°56'00'' = 341°04'00''$$

Therefore, the coordinates of point B are

$$x_\text{B} = x_\text{A} + D\sin\alpha$$
$$= 310.00\text{ ft} + (130.65\text{ ft})\sin 341°04'00''$$
$$= 267.61\text{ ft}$$
$$y_\text{B} = y_\text{A} + D\cos\alpha$$
$$= 275.00\text{ ft} + (130.65\text{ ft})\cos 341°04'00''$$
$$= 398.58\text{ ft}$$

31 *The answer is (B).*

Sections are approximately 1 mi by 1 mi, and section 11 is always due east of section 10.

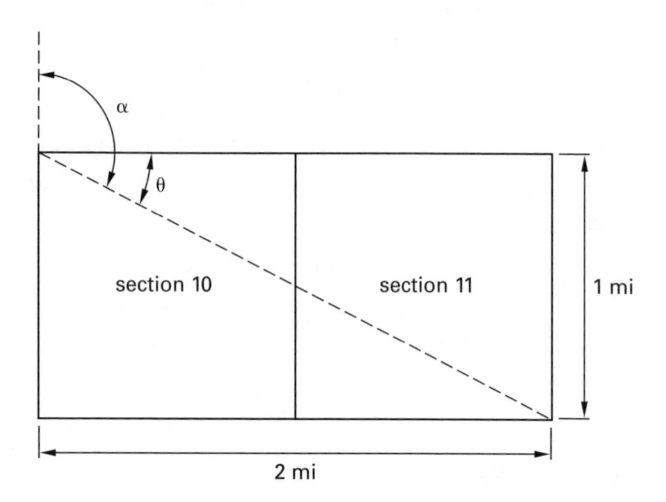

$$\theta = \tan^{-1}\frac{1\text{ mi}}{2\text{ mi}} = \tan^{-1}0.5$$
$$= 26°33'54''$$

The azimuth is

$$\alpha = 90° + 26°33'54''$$
$$= 116°33'54''$$

32 *The answer is (C).*

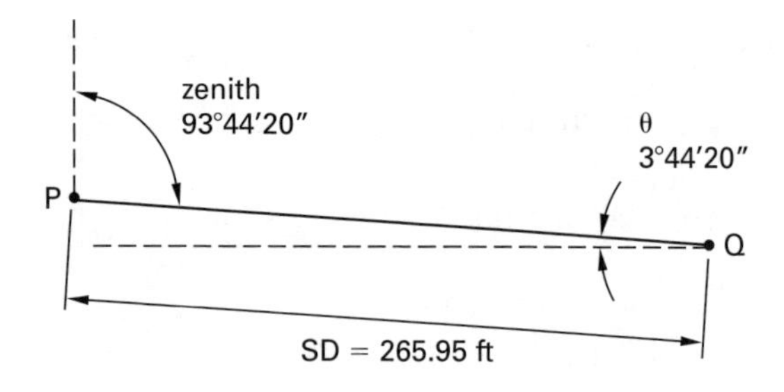

The vertical angle, θ, is

$$\theta = 90° - \text{zenith angle}$$
$$= 90° - 93°44'20''$$
$$= -3°44'20''$$

Line PQ slopes down. Therefore, the horizontal distance is

$$\text{HD} = \text{SD}\cos\theta$$
$$= (265.95\text{ ft})\cos -3°44'20''$$
$$= 265.38\text{ ft}$$

33 *The answer is (C).*

Determine the height of the instrument.

$$\text{HI} = y_\text{A} + \text{BS}_\text{A}$$
$$= 100.00\text{ ft} + 4.87\text{ ft}$$
$$= 104.87\text{ ft}$$

Determine the elevation of point D by taking the difference of the height of instrument and intermediate foresight.

$$y_\text{D} = \text{HI} - \text{IFS}_\text{D}$$
$$= 104.87\text{ ft} - 4.77\text{ ft}$$
$$= 100.10\text{ ft}$$

34 *The answer is (C).*

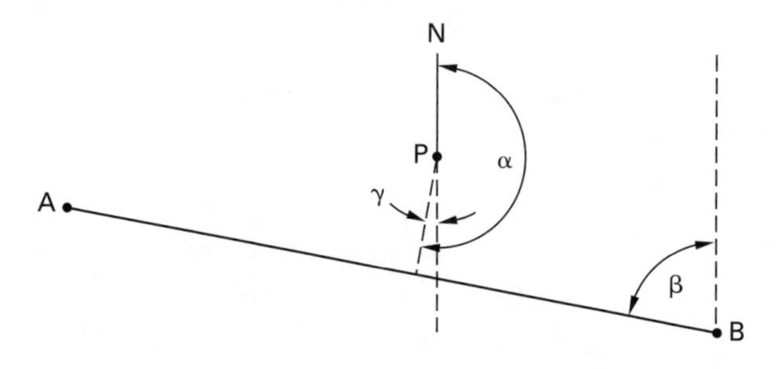

β is 79°34′50″, since bearing BA is N 79°34′50″ W. Therefore,

$$\gamma = 90° - \beta$$
$$= 90° - 79°34'50''$$
$$= 10°25'10''$$

The azimuth of the shortest line from point P to line AB is

$$\alpha = 180°00'00'' + \gamma$$
$$= 180°00'00'' + 10°25'10''$$
$$= 190°25'10''$$

35 *The answer is (C).*

A section is normally 1 mi by 1 mi in extent. A mile, or 5280 ft, is equal to 80 chains (a chain being 66 ft in length). One square mile (the area of a section) is thus 80 chains by 80 chains, or 6400 square chains. Since 10 square chains equals 1 ac, the area of a section is 640 ac.

36 *The answer is (D).*

The two projections used for state plane systems are the transverse Mercator and the Lambert conformal conic. The latter is used in California.

37 *The answer is (D).*

All of the above could be used to compute the basin volume. However, method (D) is preferred, since areas of contour shapes can be quickly determined by x-y digitizing or from planimeter measurements. The end-area formula then yields the volume.

38 *The answer is (C).*

The overlap is 60% and, therefore, the distance between the photo centers is 100% – 60%, or 40%. Since the photo is 9 in wide, this represents 40% of 9 in, or 3.6 in.

The photo scale 1 in:400 ft means that 1 in on the photo represents 400 ft on the ground. Therefore, 3.6 in on the photo represents (3.6)(400 ft), or 1440 ft.

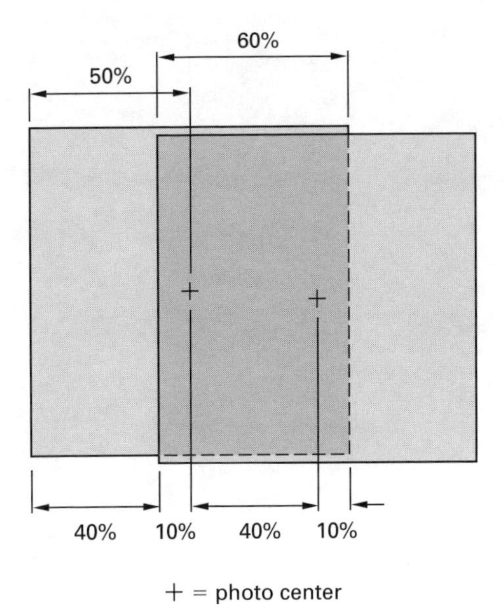

39 *The answer is (B).*

A final map, also known as a subdivision map, is defined in the Subdivision Map Act as a map that is required when a piece of land is subdivided into five or more smaller parcels.

40 *The answer is (B).*

First, determine the latitude and departure from point A to point C.

$$\begin{aligned}\text{lat}_{\text{AC}} &= \text{lat}_{\text{AB}} + \text{lat}_{\text{BC}}\\ &= 78.34 \text{ ft} - 176.38 \text{ ft}\\ &= -98.04 \text{ ft}\end{aligned}$$

$$\begin{aligned}\text{dep}_{\text{AC}} &= \text{dep}_{\text{AB}} + \text{dep}_{\text{BC}}\\ &= 64.90 \text{ ft} + 94.67 \text{ ft}\\ &= 159.57 \text{ ft}\end{aligned}$$

Distance AC can then be obtained from

$$\begin{aligned}D_{\text{AC}} &= \sqrt{(\text{lat}_{\text{AC}})^2 + (\text{dep}_{\text{AC}})^2}\\ &= \sqrt{(-98.04 \text{ ft})^2 + (159.57 \text{ ft})^2}\\ &= 187.28 \text{ ft} \quad (190 \text{ ft})\end{aligned}$$

41 *The answer is (A).*

The acronym POB means "point of beginning" and is frequently used in legal descriptions that describe the boundaries of parcels of land.

42 *The answer is (B).*

A digital terrain model is a digital map that includes contours and all vector detail, such as road centerlines and sidewalks.

A triangulated irregular network (TIN) file is a set of points for which the northing, easting, and elevation are known, and which has been ordered into a series of triplets so that all triplets form a continuous terrain surface.

A triangulated network is another name for a TIN file.

A digital elevation model is a series of northing, easting, and elevation points that defines the terrain surface. It is the modern digital alternative to a contour map.

43 *The answer is (C).*

NAD datums are horizontal datums and are not related to vertical measurements. NGVD 88 is a vertical datum used post-1991. The datum used prior to 1991 is NGVD 29.

44 *The answer is (B).*

All of the given overlap percentages are used in photogrammetry, but 60% is only used (infrequently) on projects involving digital orthophotos or photo maps. Side overlaps are almost always 30%.

45 *The answer is (D).*

The Subdivision Map Act states that when a property is subdivided into four or fewer parcels, a parcel map is required.

46 *The answer is (A).*

The opposite midpoints of a section's sides divide the section into quarters. Therefore, these points are called quarter corners.

47 *The answer is (B).*

The height of instrument measured with a tape would only give the height of the instrument above the ground at the instrument. If there were a new instrument set up between each point measured on the section, the height of instrument would not be required, but this is never the case in the field. By adding the rod reading to the known elevation of a backsight, the height of instrument above a datum can be computed.

48 *The answer is (D).*

The universal method of checking the computation of elevations from differential leveling is to subtract the sum of the backsights from the sum of the foresights, and then to compare this value with the elevation difference between the first and last point in the level route. Option (C) is in a sense a check, but is not as effective as the method just described. Option (B) is not a check, since it compares an absolute value with differences. Option (A) is almost the same as option (C), but again is not as effective as option (D).

49 *The answer is (B).*

Although the unit of measure in public land surveys today is the U.S. survey foot, the original unit was the chain, with a length of 66 ft.

50 *The answer is (A).*

The exact conversion factor between the SI unit meters and the U.S. survey foot is 3937/1200. Option (B) is a good approximation, and option (C) converts from meters to customary U.S. feet. The factor 25.4 given in option (D) is 304.8/12 (a conversion between millimeters and inches) and is approximate when applied to the U.S. survey foot.